Work-Life Advantage

RGS-IBG Book Series

For further information about the series and a full list of published and forthcoming titles please visit www.rgsbookseries.com

Published

Work-Life Advantage

*Sustaining Regional Learning
and Innovation*

Al James

WILEY Blackwell

Registered Office(s)
John Wiley & Sons, Inc., 111 River Street, Hoboken, NJ 07030, USA
John Wiley & Sons Ltd, The Atrium, Southern Gate, Chichester, West Sussex, PO19 8SQ, UK

Editorial Office
9600 Garsington Road, Oxford, OX4 2DQ, UK

For details of our global editorial offices, customer services, and more information about Wiley products visit us at www.wiley.com.

Wiley also publishes its books in a variety of electronic formats and by print-on-demand. Some content that appears in standard print versions of this book may not be available in other formats.

The information, practices and views in this book are those of the author(s) and do not necessarily reflect the opinion of the Royal Geographical Society (with IBG).

Library of Congress Cataloging-in-Publication Data

Names: James, Al, author.
Title: Work-life advantage : sustaining regional learning and innovation / Al James.
Description: Hoboken : Wiley-Blackwell, 2017. | Series: RGS-IBG book series |
   Includes bibliographical references and index.
Identifiers: LCCN 2017026665 (print) | LCCN 2017040536 (ebook) |
   ISBN 9781118944820 (pdf) | ISBN 9781118944813 (epub) |
   ISBN 9781118944844 (hardback) | ISBN 9781118944837 (paper)
Subjects: LCSH: Flextime. | Women–Employment–Psychological aspects. |
   Work and family. | Organizational learning. | Organizational change. |
   BISAC: SCIENCE / Earth Sciences / Geography.
Classification: LCC HD5109 (ebook) | LCC HD5109 .J36 2017 (print) | DDC 306.3/6–dc23
LC record available at https://lccn.loc.gov/2017026665

Cover Image: Photograph © Al James, 2016
Cover Design: Wiley

Set in 10/12pt Plantin by SPi Global, Pondicherry, India
Printed and bound in Malaysia by Vivar Printing Sdn Bhd

10 9 8 7 6 5 4 3 2 1

For friends and colleagues at QMUL Geography
(who always enjoyed a good bottle of workahol)

Contents

List of Figures

List of Tables

Series Editor's Preface

The RGS-IBG Book Series only publishes work of the highest international standing. Its emphasis is on distinctive new developments in human and physical geography, although it is also open to contributions from cognate disciplines whose interests overlap with those of geographers. The Series places strong emphasis on theoretically informed and empirically strong texts. Reflecting the vibrant and diverse theoretical and empirical agendas that characterise the contemporary discipline, contributions are expected to inform, challenge and stimulate the reader. Overall, the RGS-IBG Book Series seeks to promote scholarly publications that leave an intellectual mark and change the way readers think about particular issues, methods or theories.

For details on how to submit a proposal please visit:
www.rgsbookseries.com

David Featherstone
University of Glasgow, UK
RGS-IBG Book Series Editor

Preface and Acknowledgements

This book is about the everyday struggles of knowledge workers to juggle competing activities of paid work, home and family. It is about making visible the gendered networks of social reproduction and household divisions of labour which unavoidably shape the differential abilities of workers to perform as 'human capital' inputs to firms' knowledge production processes on a daily basis. It is about demonstrating the considerable benefits that can accrue to employers if they are willing to provide 'alternative' working arrangements for workers, to help them and their families achieve a better work-life balance. And it is about exploring the possibilities for more socially inclusive forms of regional learning, innovation and growth, and challenging the marginalisation of workers with significant caring responsibilities and personal life commitments.

Set against the backdrop of an increasingly abstract, self-referential and firm-centric regional learning and innovation research agenda, the book examines the everyday work-lives of over 350 information technologists in two high-profile high-tech clusters (Dublin and Cambridge). It documents how the everyday workplace practices so widely celebrated by economic geographers as supporting regional learning and innovation can also be socially damaging, with multiple negative outcomes for workers' careers, health, well-being and quality of life. In response, it also identifies the kinds of employer-provided work-life balance arrangements that different groups of workers and their families find most useful in seeking to reduce those work-life conflicts – and the ways that those arrangements can simultaneously enhance firms' capacities for learning and innovation, in pursuit of long-term sustainable competitive advantage. Ultimately, the analysis exposes and disrupts a series of taken-for-granted assumptions and masculinist economic universals within economic geography's flagship regional learning and innovation literature. This includes the role of gendered work-life conflict and uneven work-life provision in motivating and constraining the cross-firm job-to-job mobility of workers and the skills they embody – this in a manner that makes those knowledge spillovers much more complex than economic geographers have previously been able (and indeed willing) to recognise. It also explores the spatial variability of these high-tech work-lives and gendered learning dynamics within and between different regional economies as a function of different urban infrastructures of care and national welfare regimes.

This book began in 2006 and, 11 years on after a couple of brief intermissions, owes a lot to a lot of people. I want to thank the 350 plus technologists, programmers, software architects, CEOs, managers, network specialists, marketing professionals, HR managers and industry watchers who took time out from their busy lives to take part in this research through the interviews, surveys and other interactions in Ireland and the UK – but who necessarily remain anonymous in the analysis. Particular thanks also to Noreen Fitzpatrick at the Irish Work-Life Balance Network, Larry Bond at the Irish Equality Authority, Damien Thomas at Ireland's National Centre for Partnership and Performance, Paul Butler at Nexus Research, Barbara Keogh at the Irish Congress of Trade Unions, Eileen Drew at Trinity College Dublin, Rosheen Callender at the Services, Industrial, Professional and Technical Union (SIPTU), Karlin Lillington at *The Irish Times*, Mary Doolley at the Irish National Framework Committee on Work-Life Balance, Sarah Blow and Nicole Mathison at GirlGeek Dinners, and Maggie Berry at Women in Technology. All of them had a major hand in helping me to get the fieldwork up and running, and in shaping the direction and scope of the earliest phases of the analysis. And later down the line, thanks to Kerry Cable at BusinessFriend for undertaking all the interview transcription, and Martina O'Callaghan in the Labour Market Analysis Section, Ireland Central Statistics Office, for generating special Quarterly National Household Survey (QNHS) data extracts on the demographic and household situations of the IT workforce in Ireland.

This research received funding from the UK's Economic and Social Research Council (RES-000-22-1574-A), and was also affiliated to the ESRC Gender Equality Network (GeNet, RES-225-25-2001) led by Jackie Scott. I am grateful to geography colleagues at Cambridge, Queen Mary University of London (QMUL) and Newcastle for providing the encouragement/support/ sabbatical space to write this. Special thanks also to Diane Perrons. During the Irish fieldwork I was kindly hosted by Trinity College Dublin. And during the writing up stage I was hosted in Geography and Economic History at Umeå University, Sweden.

These ideas also evolved through interactions with successive cohorts of students at Cambridge, QMUL and Newcastle who took the Working in the New Economy, Spaces of Uneven Development and Geographies of Working Lives modules. Invited seminars were given at the Universities of Nottingham, Limerick, Umeå, Birmingham, Bristol, Stavanger, Turku, Glasgow, QMUL and the Institute for Education, alongside papers at various annual conferences of the American Association of Geographers, Irish Geographers, Nordic Geographers, Royal Geographical Society with the Institute of British Geographers, and the European Colloquium on Culture, Creativity and Economy. I would like to thank audiences at all those events for critical comment and encouragement on earlier versions of the final analysis presented here. Thanks also to Neil Coe and then Dave Featherstone as RGS-IBG book series editors, the book series editorial team and two anonymous reviewers (although I think I know who you are!).

The author gratefully acknowledges permissions to reuse and extend ideas contained in three previously published single-authored papers: *Journal of Economic Geography* (Oxford University Press, 2014), 14(3): 483–510; *Gender, Work and Organization* (Wiley, 2014), 21(3): 273–294; and *Gender, Place and Culture* (Taylor and Francis, 2011), 18(4/5): 655–684. Permission to reproduce Tim O'Brien's cartoon in Chapter 2 was obtained from CartoonStock.com. Every effort has been made to trace copyright holders, and we apologise for any errors or omissions in these acknowledgements.

Al James

Forest Gate, East London

March 2017

List of Abbreviations

BIS	UK Department for Business Innovation and Skills
BLS	US Bureau of Labor Statistics
CBI	Confederation of British Industry
CEO	Chief Executive Officer
CSO	Central Statistics Office Ireland
CSS	Cascading Style Sheets (document mark-up language)
CTO	Chief Technology Officer
DfEE	UK Department for Education and Employment
DTI	UK Department for Trade and Industry (subsequently BIS from 2007)
DWP	UK Department for Work and Pensions
EEDA	East of England Development Agency (ended March 2012)
EHRC	Equality and Human Rights Commission (UK)
EMEA	Europe, Middle East and Africa (business/market areas)
ESRC	Economic and Social Research Council (UK)
ESRI	Economic and Social Research Institute (Ireland)
EU-OSHA	European Union information agency for occupational safety and health
F2F	Face-to-face
FTE	Full-time employee
IBEC	Irish Business and Employers' Confederation
ICT	Information and communications technology
ICTU	Irish Congress of Trade Unions
IER	Institute for Employment Research
ISA	Irish Software Association
IT	Information technology
LLM	Local labour market
LMI	Labour market intermediary
MD	Managing Director

MIS	Management information system
MNC	Multinational corporation
MSI	Microsoft installer package file format used by Windows
NAICS	North American Industry Classification System
NCPP	National Centre for Partnership and Performance (NCPP) Ireland
NEG	New economic geography
QA	Quality assurance
SET	Science, engineering and technology
SIC	Standard Industrial Classification
SIPTU	Services, Industrial, Professional and Technical Union (Ireland)
SMEs	Small and medium-sized enterprises
STEM	Science, technology, engineering and mathematics
TUC	Trades Union Congress (UK)
VPN	Virtual private network
WERS	Workplace Employee Relations Survey (UK)
WITI	Women In Technology International
WITS	Women In Technology and Science
WLB	Work-life balance

Chapter One
Inclusive Regional Learning?

Introduction

> The fruits of a rapidly growing economy based on innovations and hard work are patently obvious. Less obvious are the costs absorbed by individuals as they take on the attributes required to succeed … Sustaining the new economy means building a new set of social institutions to support it. (Carnoy 2002: x)

> 'It's important that people talk about these work-life challenges. There are a lot of women, like myself, who are just so busy getting on with it, just so busy just trying to stay quiet. This type of material, it needs to be fed back, people need to stop and think about it'. Software Business Development Manager, female, two young children, 3-day work week, IT MNC, Dublin

Over the last two decades, the shifting spatial and temporal boundaries between work, home and family that have accompanied the transition to the so-called 'new economy' have been hotly debated. As firms reorganise work in response to globalisation and new technological opportunities, 'flexibility' for many workers has come to mean increased workloads, less predictable work schedules and more unsocial work hours as firms demand they work longer and harder to minimise labour costs. Simultaneously, household life has also become more complex as female labourforce participation rates continue to grow and an ever-increasing proportion of workers are part of dual-earner households. These problems are reinforced by the decline of the extended family, increasing lone-parent households and greater eldercare responsibilities through increased life expectancy.

Work-Life Advantage: Sustaining Regional Learning and Innovation, First Edition. Al James.

Simultaneously, the neoliberal attack on social provisioning has transferred the burden of care down to the 'natural' level of home (Bakker and Gill 2003) where most women retain the major responsibility for the 'messy and fleshy' components of domestic and family life (Katz 2001; Crompton and Brockmann 2006). The overall result is a complex, gendered, multi-variable balancing act between the competing demands of paid work and responsibilities, commitments and life interests beyond the workplace, for which workers have only 'finite resources in terms of time and energy' (Cooper et al. 2001: 50).

In response, the desirability and means of achieving an appropriate 'work-life balance' (WLB) has received widespread attention from governments, managers, trade unions, academics and the media. At the individual level, WLB refers to 'the absence of unacceptable levels of conflict between work and non-work demands' (Greenblatt 2002: 179). While encompassing earlier family-friendly perspectives, the work-life balance term was intended to broaden the debate beyond working mothers to include all workers, and hence a wider diversity of personal life needs, interests and responsibilities such as religious attendance, sports, hobbies, and community and charity work. Alternative WLB monikers include work-life reconciliation, work-personal life integration, work-personal life harmonisation and work-life articulation. But whatever the label used, the societal and moral significance of the successful integration of paid work with other meaningful parts of life is profound. Study after study has documented how a lack of work-life balance can result in increased stress, deleterious effects on psychological and physical well-being, and increased family and marital tensions (e.g. Burchell et al. 1999, 2002; Frone et al. 1994; Lewis and Cooper 1999; Scase and Scales 1998). Moreover, given persistent gender variations in work-life stress as women make the greatest compromises to fit paid work around family (Moen 2003; McDowell et al. 2005), studies have also highlighted the importance of work-life provision by employers as a means for improving gender equity in market employment and household caring (Wise and Bond 2003; World Economic Forum 2005). The labour movement has also emphasised the social importance of WLB as a means of improving workers' quality of life and combating the increasing work pressures that are destabilising households and societal integration.

Employer-provided WLB arrangements are typically split across four categories, in terms of those providing workers with greater temporal flexibility of work, greater spatial flexibility of work, reduced total work hours and childcare assistance. But despite government efforts, evidence of progress in employers providing comprehensive suites of work-life arrangements remains uneven, resulting in continuing hardship for many workers and their families. Indeed, these problems have also been exacerbated in the aftermath of the 'global' economic downturn which created new gendered work-life demands through rapid and dramatic labour market change, heightened fears of job loss, increased workloads and understaffing (e.g. Fawcett Society 2009; TUC 2009). With employers keen to effect cost savings, workplace arrangements designed to help reconcile workers' competing commitments around work, home and family have not been immune (Galinksy and Bond 2009). At the heart of this disjuncture, many scholars argue that employers are simply unlikely to implement meaningful WLB arrangements unless they can identify bottom-line economic advantages that arise from doing so (e.g. Healy 2004; Hyman and Summers 2004; Dex and Scheibl 1999; Dex and Smith 2002). Importantly, this 'WLB business case' also lies at the heart of UK, Irish and US government policy interventions in this area, with employer benefits from WLB provision widely touted by policy-makers as improved recruitment, retention, morale and productivity, and reduced stress, absenteeism and costs. Yet despite its popularity, there remains a relative dearth of empirical evidence to support these claims in practice (Beauregard and Henry 2009). In addition, 'few scholars have demonstrated the mechanisms through which such [WLB] policies function (or do not) to enhance firm performance' (Eaton 2003: 145–146).

Work-Life Advantage takes issue with this major knowledge gap and its negative social consequences for workers and their families, whose collective labours are ultimately responsible for (re)producing and sustaining some of the world's most high-profile high-tech regional economies. In so doing, the book develops a new analytical approach that connects the burgeoning research agenda on gendered labour geographies of work-life balance, social reproduction and care with an equally expansive research agenda on regional learning and innovation. Importantly, both agendas ultimately respond to the emergence of 'flexible' pro-duction processes in the wake of Fordism from the late 1970s onwards: one then exploring the territorial forms of flexible production (firm-centric focus), and the other, dramatic changes in the organisation of flexible paid work and working times as experienced by workers and their families (workerist focus). Yet despite these common roots, these two research agendas remain oddly disconnected. In seeking to bridge them, the hybrid analysis developed in this book answers four major research questions. What are the common, everyday experiences and outcomes of gendered work-life conflict amongst knowledge workers and their families in high-tech regional economies? What kinds of employer-provided WLB arrangements do different cohorts of knowledge workers find most useful in over-coming those conflicts? How does the uptake of these worker-preferred WLB arrangements enhance (vs. constrain) the kinds of intra-firm and cross-firm learning and innovation processes widely identified as enabling regional advantage? And do those WLB learning outcomes vary both within and between regional economies, particularly as a function of national welfare regimes? In so doing, the book responds to earlier calls by Lewis et al. (2003) to develop a 'dual agenda' that moves beyond either/or thinking to consider *both* business and social imperatives in pursuit of optimal work-life balance outcomes, set within a regional learning framework.

This analysis is developed through a case study of information technology (IT) workers and firms in Dublin, Ireland and Cambridge, UK prior to and after the onset of the Great Recession in 2008. Crucially, both regions have figured prominently in regional learning and innovation studies to date, and are recognised as important European clusters of IT growth of interest to policy-makers elsewhere. Additionally, IT represents a knowledge-intensive industry at the vanguard of new working practices, in which firms compete to bring new products to market quickest and in which 'work' and 'life' are significantly blurred. Work-life balance has also come to assume a strong national significance in both Ireland and the UK, as a function of long average work hours relative to other EU member states. The book builds on 10 years of research, including an ESRC-funded research project (2006–2009): *The Impacts of Work-Life (Im)Balance on Innovation and Learning in Regional Economies* (RES-000-22-1574-A). Its critical analysis draws on a rich, multi-method evidence base comprising two regional surveys of IT employers (150 firms with combined local employment of 8,068 workers); 68 in-depth interviews with female and male IT workers, HR managers and labour organisers; and a WLB/labour mobility survey of 162 female IT workers (conducted through a number of women's IT networking organisations: Girl Geek Dinners, Women In Technology, WITS Ireland). It also draws on recessionary data from a second survey of 139 female IT workers conducted in December 2010.

This introductory chapter continues with some vivid examples of the daily realities of gendered conflicts between work, home and family that I have documented amongst IT workers in the UK and Irish contexts, with varying levels of employer support. These examples are juxtaposed against an increasingly abstract and firm-centric regional learning and innovation literature in which workers are treated 'not as social agents capable of making landscapes in their own right but, rather, as simply an aspect of capital' (Herod 2001: 22). Consequently, these studies have paid almost zero attention to how workers' gendered identities, varied responsibilities of care and personal-life interests beyond the workplace unavoidably shape their (non-)participation in the relational networks and communities of practice widely theorised as enabling regional learning and innovation.

This chapter also outlines the book's wider contribution to an enhanced interdisciplinary conversation between economic, labour and feminist geography, as part of an 'intellectual trading zone' (Barnes and Sheppard 2010) to develop richer, pluralistic understandings of how regional economies function – explored through the everyday work-lives of the engineers, scientists and technologists whose collective labours are ultimately responsible for (re)producing and sustaining them. The wider policy significance of this push for 'engaged pluralism' is also introduced as part of a holistic regional development agenda (Pike et al. 2006), which integrates mainstream economic concerns around competitiveness, growth and productivity with normative questions around labour market inclusion, gender equity, worker well-being and social reproduction (see also Rees 2000; Perrons 2001; Blake and Hanson 2005). The final section sets out the structure of the book on a chapter-by-chapter basis.

Unlearning Regional Learning? (Or why economic geographers need to get out more)

It is now widely accepted that fundamental changes within advanced capitalist economies since the 1970s herald a new era of capitalist economic development, whose geographical form is marked by a decisive reagglomeration of production and the rise of regions as the salient foci of wealth creation (Martin and Sunley 2003). Characteristic in their high rates of technological learning and innovation,[1] the workings of these 'industrial agglomerations' or 'clusters' have become a fixation for successive cohorts of economic geographers ever since. Newcomers to this literature are now confronted with an overwhelming plethora of territorial innovation models and concepts used to unpack the learning advantages of 'being there' (Gertler 2003). In a nutshell, this 'regional advantage' (Saxenian 1994) is understood as emergent from enhanced access to external information flows and knowledge spillovers between co-located firms, research organisations and public agencies; rooted in networks of repeated face-to-face interaction, the cross-firm mobility of talent and through processes of new firm formation and spin-off. Geographers have also analysed differences in firms' 'absorptive capacities' (Cohen and Levinthal 1990), understood as the distinctive sets of 'shared' socio-cultural norms and conventions upon which actors routinely draw to assimilate and apply externally derived information and knowledge within the firm to develop new products or services, new technological capabilities and/or new ways of organising production processes and service delivery.

Without doubt, this literature has yielded important insights into the socio-cultural foundations of regional economic advantage, fundamentally shaped the content and focus of regional economic development policies worldwide and become a major cornerstone of economic geography as a sub-discipline. Yet (as I explore in Chapter 2), the international research agenda around regional learning and innovation continues to suffer from a series of peculiar blindspots, emergent from: reductionist treatments of labour as passive inputs to firms' innovation activities; repeated analytical exclusions of female workers; and an overriding tendency to artificially divorce networks of knowledge production from extra-firm networks of social reproduction. And all this despite a high-profile call-to-arms over a decade ago:

> Neither of the NEGs [New Economic Geographies] pays any attention to questions in the immediate sense of the social division of labour between different kinds of paid work and between paid work and caring, or the wider sense of establishing sustainable regional development. Yet these dimensions are central to understanding the well-being of people within regions and therefore to regional or spatial development as a whole. (Perrons 2001: 211)

The societal significance of this peculiarly narrow focus of the regional learning agenda is poignantly illustrated in the following quotes, which juxtapose some core tenets of the regional learning literature and its simplistic treatment of labour as 'knowledge agents' with personal testimonies of the everyday lived realities of doing knowledge work by IT workers in Dublin and Cambridge, two of Europe's leading high-tech clusters:

> The active units behind the formation of new knowledge are 'epistemic communities', simply defined as groups of knowledge-driven agents linked together by a common goal, a common cognitive framework and a shared understanding of their work. (Cohendet et al. 2014: 930)

> 'I'm the CEO of [IT company] and I'm also the mum of two kids. So as an employer, I'm thinking about how I want to create an environment for the people who work in my company, and also honour the commitments I have to my shareholders. Pretty much the stress comes from wanting to be successful at work, and also wanting to be successful as a mother, or wanting to be successful at a hobby, or wanting to do a lot of different things and having the conflict'. CEO, female, IT start-up, UK SE region

> Agglomeration does not ensure learning or determine its content. [Rather] the use and development of information in such a way that technological learning takes place has to do with the qualitative behaviours of agents in a network. (Storper 1997: 135)

> 'Your life is lived at a pace, it's lived at a speed ... you actually don't know what it is to relax ... so in 20 years' time what will we all have remembered of this period in our lives? What will the children remember? What will their experience of being a child in our home be? What will my personal experience be? What will [my husband]'s experience be? ... There will be a lot more money in the bank, that's if we're still alive, if we're all still talking to each other, if, you know, the kids haven't gone off the rails because we haven't had time to sit down and talk because there hasn't been proper family conversation.' Software Business Development Manager, female, two young children, 3-day work week, MNC, Dublin

These examples of the daily realities of juggling work, home and family are drawn from workers who populate the same high-tech knowledge-intensive sectors and regional industrial clusters that have long formed the substantive focus of the regional learning literature. They raise important questions concerning the wider social impacts of current modes of time-based competition and high-tech regional development on workers and their families. But these questions remain unanswered (if indeed they are even posable at all) within an increasingly abstract regional learning and innovation literature which offers minimal engagement with the critical labour geographies agenda instead concerned 'to see the making of the economic geography of capitalism through the eyes of labour' (Herod 1997: 30; see also Castree 2007; Lier 2007; Coe and Lier 2011; Rutherford 2010). The danger, then, is that the negative social outcomes of routine innovation and learning activities – as experienced by workers and their families – are ignored. Worse still, copycat cluster policies based on emulating other economically successful regions run the risk of reproducing those negative social outcomes and work-life conflicts elsewhere (see also Rees 2000).

The core argument here, then, is that we need to disrupt or 'unlearn' a series of widely held 'economic universals' (Gibson-Graham 2006) within the regional learning literature, rooted in genderless conceptions of 'human capital' as factor inputs to firms' knowledge production and exchange processes. These include the masculinist myth of the disembodied 'ideal worker' for whom work is primary and the demands of family and personal life insignificant; taken-for-granted assumptions regarding the rootedness of cross-firm knowledge spillovers in apparently 'boundaryless careers' and a singular commitment to advancing technological innovation above all else; and the (falsely) assumed autonomy of regional 'innovative milieux' from regional 'social *re*productive milieux'. The broader project is long overdue, motivated by the recognition that 'the full complexity of modern economies only becomes apparent when we move outside what are often still considered to be the "normal" territories of economic inquiry. Then a whole new world hoves into view' (Thrift and Olds 1996: 311).

Extending the Feminist/Labour/Economic Geography Trading Zone

Work-Life Advantage connects ongoing debates in economic geography around the regional foundations of learning and innovation to a high-profile set of feminist and labour debates around the shifting boundaries between work, home and family, and the stubborn resilience of gender inequalities in paid work and caring. In so doing, it responds to recent calls for a greater level of 'engaged pluralism' across multiple sub-disciplinary communities of practice and cultures of inquiry within human geography (Barnes and Sheppard 2010), through the development of new intellectual 'trading zones' based on mutual dialogue and productive conversation. In the process, scholars are encouraged to step outside their intellectual comfort zones and sub-disciplinary specialisms, trade ideas and theories around common problems, expose conversational partners to alternative viewpoints and critiques which prompt them to question their own position, and thereby catalyse new understandings and possibilities through give and take. Barnes and Sheppard contrast this with the 'fragmented pluralism' that they identify within contemporary economic geography rooted in 'a series of intellectual solitudes that has created isolation, producing monologues rather than conversation, and raising the question of how knowledge production should proceed' (2010: 193). The ultimate aim, then, is to develop genuinely new hybrid understandings of the complex geographies of economies, in a manner that moves away from the (re)production of economic geography 'versus' feminist geography as institutionally bounded sub-disciplines.

Crucially, it is important to recognise that a feminist/labour/economic geography trading zone *already* exists, albeit one that is partial and asymmetrical, based on feminist scholars engaging with and recasting economic and labour theory (rather than 'mainstream' economic geographers reaching out from the other direction). This trading zone is evident in relation to major advances in our understanding of the relationship between patriarchy and capitalism (e.g. McDowell 1991; Gibson-Graham 1993); gendered labour market inequalities and exclusionary masculinist workplace cultures (e.g. McDowell 1997); female labour market precarity and feminised labour organising (Hardy 2010); gendering the financial crisis and feminised financialisation (e.g. McDowell 2010; Pollard 2012); and gendered divisions of 'hidden' household reproductive labour and 'life's work' (e.g. Massey 1997; Mitchell et al. 2004). This work has generated a series of fascinating new empirical, conceptual and theoretical insights, new socio-economic imaginaries, new vocabularies, new methodological innovations, and given new analytical voice to previously marginalised actors and places. Most notably, it has disrupted conventional masculinist conceptions of work as something that is done in the public sphere outside the home, for a wage, governed by an employment contract (compare with a broader conception of work done for love, without a wage, within the home, without an employment contract, typically by women (Massey 1997)). In so doing, feminist geographers have sought to reimagine economy and to challenge male economic 'universals' at the heart of commonly accepted theories of how we think economies and labour markets function.

However, whilst recognition of the significance of gender divisions – and associated body of feminist scholarship – has the potential to transform economic geography through new dialogues, key research agendas in economic geography remain limited in their analytical engagement with gendered social relations and the ways in which they are both partly constituted by and affect economic processes (McDowell 2000). Conspicuously, the regional learning and innovation agenda does not figure prominently within this feminist/economic geography pluralist project. Many feminist geographers steer well clear of what they perceive to be an increasingly abstract, self-referential and myopic regional learning and innovation literature narrowly focused on the minutiae of firms' productive processes at the expense of broader labour concerns of gender inequality and female worker empowerment. The aim in this book, then, is to extend that trading zone to the regional learning and innovation research agenda, in order to challenge its stubborn analytical silence on female worker agency in actively (re)producing regional geographies of learning, innovation and growth in practice. It is also concerned to analyse the variety of gendering and sexing practices which shape the different terms on which female and male workers – with a variety of caring responsibilities and personal life commitments – are able to engage in the relational networks of learning and communities of practice long theorised as underpinning regional advantage. This in ways that necessarily make those regional learning dynamics much more complex than we have previously recognised (and indeed been *willing* to recognise).

For my part over the last decade, stepping into the feminist/labour/economic geography trading zone has involved critical dialogue with colleagues with overlapping interests in gendered geographies of economies (albeit with different sub-disciplinary identities, backgrounds and training).[2] These conversations have focused around the workplace and labour market exclusions of female technologists; gender, motherhood and the economic marginalisation of women; femininities in white collar workplaces; gendered health geographies of work-life conflict; feminist 'neoliberalisms'; and on situating and disrupting 'universal' theories of economy. As part of this process, I have also presented and debated my evolving research ideas around gendered regional learning at multiple forums including: the ESRC workshop series on feminism and futurity; European seminars on regional learning and innovation; the Medical Research Council seminar series on gender and public health; feminist geographies and labour geographies sessions at the annual international conferences of the RGS-IBG, AAG and Nordic Geographers Meeting; the lifelong learning seminar series at the UK's Institute for Education; and 'straight' economic geography seminars. However, this process of constructive pluralistic dialogue did not always live up to the positive and encouraging portrayal laid out by Barnes and Sheppard. Whilst a common critique from regional learning scholars was that this work is 'not economic enough' (!), from some feminist scholars I was charged with 'missing the point', on the basis that a focus on 'the WLB business case' subverts the genuine moral and ethical claims for work-life balance and inclusive workplaces.

Similar criticisms also emerged as a product of the peer review process for the *Journal of Economic Geography*, *Gender, Place and Culture* and *Gender, Work and Organisation*, with reviewer comments typically centred within (and hence perpetuating) clearly defined sub-disciplinary domains (typically regional learning *or* feminist geography, but rarely both). In this way, the journal peer review process is exposed as part of the material research practices that reproduce disciplinary boundaries, in which research activities are validated, sub-disciplinary cultures learned and 'credentialed practitioners' promoted who can speak with authority within specified disciplinary communities of practice (Schoenberger 2001: 370; Miller and Fox 2001). This is reinforced by increasingly strict journal word count limits which too often preclude wide-ranging, integrative literature reviews drawn from multiple research agendas. This sub-disciplinary boundedness was also evident in the comments of one reviewer of this book proposal, that 'the mainstream economic geographers interested in questions of regional innovation and competitiveness will not read the book', whilst also preaching to the choir in feminist geography *already* interested in questions of work-life and labour. The point of this book, then, is to bring these multiple versions of earlier papers together and to engage with a broader audience of human geographers on their own terms – this through a much deeper and integrative analysis than has previously been possible through journal publication. The luxury of 95,000 words certainly helps, coupled with a diverse editorial review team of eight colleagues drawn from multiple sub-disciplinary backgrounds across human geography who were willing to take the chance. This remains a distinctive and attractive feature of the RGS-IBG book series.

Advancing the Holistic Regional Development Agenda

> Who pays for the kids? This is the short version of a larger question: How are the costs of caring for ourselves, our children, and other dependents distributed amongst members of society? These costs are largely paid by women, both inside and outside the money economy. And they seem to be increasing. (Folbre 1994: i)

The wider policy relevance of the analysis developed in this book stems from its empirical and conceptual contributions to a growing agenda around 'holistic regional development' (Pike et al. 2006). This project is concerned with expanding the narrow analytical focus of regional studies beyond economistic – or 'dessicated' (Morgan 2004) – indicators of competitiveness, growth and productivity, also to include gender equity, quality of life and well-being amongst workers and their families. A core challenge involves identifying the *practical* and *organisational* means for promoting more socially inclusive forms of growth, in a manner that challenges the labour market marginalisation of relatively excluded, disadvantaged and vulnerable groups, in pursuit of more evenly shared prosperity. Indeed, these concerns have grown particularly in the wake of the financial crisis, economic downturn and subsequent period of austerity, with policy interest in socially inclusive growth also increasingly evident amongst national and international governments and think tanks (e.g. European Commission 2010a; OECD 2014). In sum, this international agenda seeks to foster:

> Long-term sustainable economic growth that creates economic opportunity in the form of decent and productive employment ... that may be accessed by all of society regardless of economic status, gender or ethnicity, thus enabling all of society to both directly benefit from and participate in economic activity and future growth. (Rodríguez-Pose and Wilkie 2015: 13)

As such, this progressive agenda positions the twin policy ideals of economic prosperity and social inclusion as not simply potentially reconcilable, but argues that social inequity is simply bad for economic growth (Benner and Pastor 2015). Accordingly, it emphasises the importance of enabling increased labourforce participation and labour market opportunity amongst marginalised groups, on the basis that sustained growth requires a diverse employment base. Overlapping concepts include: good growth (PWC and Demos 2013), just growth (Benner and Pastor 2012) and equitable economic growth (Rodríguez-Pose and Wilkie 2015). Regardless of the label used, the consensus is that 'other forms of growth' are possible (and needed), and that the traditional priority of first 'fixing the economy' as a prelude to and platform for securing social well-being is insufficient (Pike et al. 2006: 256).

I argue that the mainstream regional learning agenda – with its general lack of engagement with questions of gender equity, female worker agency, social well-being and the negative work-life outcomes of high-tech regional development – runs the risk of becoming marginalised from (indeed irrelevant to) an ascendant international policy agenda around socially inclusive growth. In this way, it is imperative that regional learning scholars respond to growing calls for economic geographers to develop more critical, heterodox and pluralist analyses of *how* and *where* economies function, *for whom*, and *to what ends* (Christophers et al. 2016). Previous work has explored how the pursuit of more socially inclusive modes of economic growth is dependent upon concerted action and the development of supportive forms of governance and social regulation at the local, urban, national and supra-national scales (Benner and Pastor 2012, 2015). However, less attention has been paid within inclusive growth debates to employers as institutions that differently govern the work-lives and well-being of workers and their families. And this despite the prioritization of employment accessibility over *ex post* redistribution as a means for reconciling potential tensions between economic growth and social equity, on the basis that 'increased labour force participation will support and perhaps even act as a catalyst for continued growth' (Rodríguez-Pose and Wilkie 2015: 1). Key questions remain, therefore, around the potential role of different forms of work-life balance provision by employers as viable levers for promoting socially inclusive growth and increasingly feminised labour markets in high-tech regional economies. Specifically, what is the scope for enabling *more* rather than less *accessible* forms of employment for workers with significant caring responsibilities, through the provision of 'alternative' or 'non-standard' working arrangements designed to help workers reconcile ongoing conflicts between paid work, home and family that would otherwise force them to quit?[3] Herein lies the wider relevance of the expanded regional learning agenda and intellectual trading zone outlined above, concerned to identify gendered social inequities in the work-life experiences of similarly qualified workers in doing regional learning and innovation in practice, and to consider business, labour and social imperatives in pursuit of optimal work-life balance outcomes.

Work-Life Advantage: Building the Argument

The original analysis developed over the next seven chapters of this book brings the everyday labour struggles of technology workers and their families to combine work, home and family to the fore of an expanded regional learning agenda. In the next chapter, I provide a critical introduction to the expansive regional learning and innovation research literature, through which geographers have so far advanced our theoretical understandings of (masculinist) regional advantage. These intellectual contributions are illustrated through a range of influential studies, yet whose contributions are also striking in their firm-centrism, gender myopia and/or rootedness in the myth of the disembodied 'ideal worker' (Williams 2000) for whom work is primary, time available to work unlimited, and the demands of family and personal life insignificant. In seeking to disrupt these economic orthodoxies, the chapter documents a growing body of alternative geographical research that builds on Massey's classic (1995) study of female technologists in masculinised 'high-tech monasteries' to explore female worker agency, female-dedicated cross-firm learning infrastructures and feminised high-tech communities of practice. This alongside the increasing numbers of 'Silicon Cowboys' who are hanging up their spurs and undertaking significant caring responsibilities outside of paid work. These changes in turn underscore the crucial significance of the holistic regional development agenda, with its attendant focus on work-life balance, care, quality of life and gender inclusion as vital (albeit strangely neglected) factors for reproducing regional learning and innovation in practice.

The expansive work-life balance research agenda forms the second major research stream which frames this monograph, and this is introduced in Chapter 3. Here, I explain the phenomenal rise of the WLB agenda over the last two decades; outline multiple and competing work-life terminologies; and explore the labour market, health and family consequences of work-life conflict. The chapter also examines the increasing salience, policy appeal and limits to the WLB 'business case', juxtaposed against its major conceptual/methodological limits namely: (i) prioritisation of employer needs at the expense of workers and their families; (ii) limited focus on output measures of firm performance, rather than sources of long-term sustainable competitive advantage (learning and innovation capacities); and (iii) atomisation of firms from the regional industrial systems of which they form part. These three critiques motivate the alternative regional learning/WLB 'mutual gains' approach developed in the subsequent chapters of the book.

Chapter 4 contextualises the two regional case studies (Dublin and Cambridge) and explores the methodological practicalities of researching regional geographies of work-life balance and learning in practice. The book draws on 10 years of research, including two major periods of fieldwork undertaken with IT workers and firms in Dublin, Ireland and Cambridge, UK prior to (2006–2008) and subsequent to (2010) the onset of the economic downturn. The multi-method data collection and analysis strategies are outlined, with the latter connecting managers' and workers' perceptions and lived experiences of the learning benefits of WLB provision within their respective firms, with measured changes in the performance of those same firms across multiple metrics over the same time period.

Based on these data, Chapter 5 explores the multiple negative outcomes of work-life conflict experienced by IT workers and their families – this as they seek to grapple with the wider social consequences of time-based competition, in ways previously undocumented in regional learning analyses. In addition to compromising workers' health, well-being and quality of life, the analysis also identifies negative outcomes for firm competitiveness through employees under-performing through stress and a perceived lack of employer support; female workers often taking compromise jobs which reduce work strain but ultimately under-utilise their skill-sets, knowledge and experience; and some women quitting IT employment altogether. In response to those difficulties, the chapter also explores IT workers' preferred WLB arrangements to reduce those tensions (mutual gains Part I), compared with the kinds of WLB arrangements most commonly provided by their IT employers. Chapter 5 also identifies a series of varied changes in IT workers' experiences of work-life conflict through the recession, set against wider claims that, in the wake of the economic downturn and job losses, calls for enhancing work-life provision may seem a 'little indulgent' (EHRC 2009: 6).

Chapter 6 explores how, by making available the kinds of WLB arrangements identified by workers in Chapter 5 as offering meaningful reductions in gendered work-life conflicts, employers can also enhance capacities for learning and innovation within the firm (mutual gains Part II). Over half the managers responding to the employer survey (N = 150) indicated 'an improved corporate environment for learning and creativity' as a result of their total bundles of WLB provision, perceptions that were also consistent with measured improvements in firm performance over the same timeframe. Chapter 6 delineates three sets of causal mechanisms which underpin these patterns, centred on: (i) increased self-determination of temporal pattern and spatial location of work; (ii) increased work team heterogeneity and enhanced repertoires of competencies and networks of external contacts; and (iii) increased sustainability of learning through worker engagement. Chapter 6 also identifies some constraints on firms' learning capacities that arise from non-traditional working arrangements (e.g. around practices of agile software development), and critically examines a variety of home-working and communication technologies used by some firms to reduce them. However, for some workers these technologies are also revealed to be a double-edged sword, blurring further the spatial and temporal boundaries between work and home and generating new work-life conflicts.

In Chapter 7 I extend the intra-firm analysis developed in Chapter 6 to explore the connections between work-life balance and *cross-firm* learning networks through worker (im)mobility. The analysis reveals a darker side to the kinds of worker mobility so widely celebrated in the regional learning literature – and in regional development policy – as accelerating the transfer of embodied knowledge, expertise and technological capabilities between firms. In sum, these patterns of mobility and knowledge spillovers are also premised on gendered dissatisfactions with work-life conflict; unequal divisions of household labour; uneven and often inadequate employer WLB provision (formal arrangements and informal support practices); and worker concerns that stretch far beyond 'technological innovation' around issues of care and improved quality of life. Enhanced WLB provision increases firms' abilities to attract talented workers relative to firms with lower levels of provision. But the analysis also shows how work-life concerns *constrain* cross-firm job-to-job mobility and knowledge spillovers, as some female IT workers seek to avoid the potentially disruptive impacts of a new job on delicately balanced daily trajectories between different urban sites of paid work and social reproductive care. In so doing, the analysis questions the widely espoused, firm-centric, regional learning mantra that worker mobility is always and everywhere a good thing.

In closing the book, Chapter 8 returns to the two major research literatures whose limitations framed the alternative analysis developed here, and summarises its core arguments, major findings, and wider intellectual and policy significance. It highlights the particular irony of employers rolling back work-life provision in pursuit of short-term cost savings (see e.g. Kossek et al. 2011), which not only perpetuates significant hardships for workers and their families, but also risks removing the very same 'alternative' working arrangements which this book has identified as positively underpinning firms' learning and innovative capacities, and long-term sustainable competitive advantage. Chapter 8 also identifies several future research possibilities which emerge from this research. They include shifting the dominant focus of the WLB literature beyond how best to flexibilise work and employment around an assumed female majority responsibility for childcare, to explore how best to support men who are 'going against the grain' (Ranson 2010) in assuming a greater proportion of childcare. Likewise, extending the analysis to workers and firms in the Global South, in order that we begin to provincialise 'universal' notions of the WLB metaphor which originated in the Western context (Lewis et al. 2007), and thereby give voice to a wider variety of situated WLB experiences and knowledges. In so doing, I am keen to encourage and enable younger colleagues to extend this exciting research agenda.

Notes

1   Learning is understood as 'the process by which firms, regions, industries and countries absorb information, develop knowledge, and use this information and knowledge to improve their economic performance'; and innovation as 'new creations of economic significance of various types, including new products and services, new technological capacities, and new ways of organizing production processes and the delivery of services' (Benner 2003: 1811).

2   These conversation partners included: Mia Gray around the workplace and labour market exclusions of female technologists; Cathy McIlwaine around gender, motherhood and the economic marginalisation of women; Kim England on femininities in white collar workplaces; Carol Emslie and Kate Hunt on gendered health

geographies of work-life conflict; Wendy Larner and Erica Pani on feminist 'neoliberalisms'; and Jane Pollard and Kavita Datta on the need to situate and disrupt 'universal' theories of economy.

3   Following Gibson-Graham (2006), the potential dangers of using the terms alternative or non-standard to describe working arrangements are that this terminology 'subordinates what it designates to the "mainstream"... [and] affirms the dominant by identifying the deviant' (p. xxii). Usefully, however, it also signals that the status quo is problematic, and that there is a push to change things, to pose a challenge to 'mainstream' work norms.

Chapter Two
Recentering Regional Learning: Beyond Masculinist Geographies of Regional Advantage

Introduction

> Recognising the significance of gender divisions – and the associated body of feminist scholarship – has the potential, indeed has begun, to transform economics and economic geography. … And yet economic geography remains uneven in its analyses of gendered social relations … and the ways in which they are both partly constituted by and affect economic processes. (McDowell 2000: 497)

This book is concerned with the everyday work-lives of the engineers, scientists and technologists whose collective labours are ultimately responsible for (re)producing and sustaining some of the world's most high-profile high-tech regional economies. While the high-tech firms and support organisations which populate these regional industrial agglomerations – or 'clusters' – have commanded widespread attention over the last three decades, their attendant labour geographies remain heavily under-researched. Building on pioneering work in California's Silicon Valley (and a subsequent swathe of other Silicon Copycat regions), an expansive and highly influential regional learning research agenda has analysed in ever finer detail how spatial agglomeration helps foster conditions suitable for enhanced knowledge creation, learning, innovation and inventiveness – that is, for sustaining firms' long-term competitiveness through the development of new ideas, to anticipate and outrun attempts at imitation by competitors (Lundvall 1992; Asheim et al. 2007). Indeed, 'geographical agglomerations, or "clusters"… have been selected in recent years by scholars from a number of different

Work-Life Advantage: Sustaining Regional Learning and Innovation, First Edition. Al James.
© 2018 John Wiley & Sons Ltd. Published 2018 by John Wiley & Sons Ltd.

disciplines as *the* territorial configuration most likely to enhance learning processes' in a world of heightened spatial competition (Maskell 2001: 922).

This chapter offers an introduction to, and critique of, this expansive regional learning and innovation literature, as the first of two major streams of research which in combination motivate the alternative workerist analysis developed in this monograph (regional learning meets work-life balance). I begin by outlining a series of key concepts and core propositions that have advanced our understanding of the geographical determinants of learning, innovation and knowledge exchange in dynamic regional economies. I argue that despite the expansive nature of this heterodox research literature and its apparent socio-cultural sensibilities, the majority of regional learning analyses to date remain strangely gender-blind, or else rooted in the myth of the disembodied 'ideal worker' (Williams 2000) for whom work is primary, time available to work unlimited, and the demands of family, personal life and social reproduction insignificant. In so doing, geographical analyses of regional learning have rendered invisible a series of everyday hardships amongst knowledge workers and their families, which emerge from their struggles to reconcile competing activities of home, work and family. As I show in this book, much more than a problem for firms' employees, these work-life conflicts also have major consequences for *corporate* capacities for learning, innovation and socio-economic competitiveness.

The path-dependent lineage of this analytical blinkeredness is explored in the second half of this chapter in relation to AnnaLee Saxenian's (1994) seminal study of Silicon Valley's pioneers: young male engineers in their early twenties, without family commitments and responsibilities. This heroic portrayal of the 'Silicon Cowboy' is contrasted with recent demographic workforce shifts in maturing SET sectors, associated policy interventions around women in technology, and a growing body of research that has begun to explore the gendered labour geographies of high-tech regional learning. Building on Massey's earlier (1995) analysis of masculinised 'high-tech monasteries', this work has begun to bring important advances in feminist geography to bear upon regional learning accounts and, in so doing, exposed the masculinist limits of 'universal' theories of regional advantage. Indeed, increasing numbers of Silicon Cowboys are *also* hanging up their spurs and undertaking significant caring responsibilities outside of paid work. Accordingly, the chapter explains the crucial significance of a socially inclusive (or holistic) regional learning agenda, with its attendant focus on female worker agency, gendered labour geographies of work-life balance and social reproductive care.

The aim, then is to bring labour geography and feminist geography to the very heart of the regional learning and innovation agenda. So motivated, the alternative analysis developed in this book makes visible the everyday struggles of knowledge workers in high-tech regional economies to combine competing activities of paid work, home and family – and the consequences of those labour struggles for the innovative capacities and competitiveness of the firms that employ them.

Theorising Regional (Dis)advantage

It is now widely accepted that fundamental changes within advanced economies since the 1970s herald a new era of capitalist economic development. Variously conceptualised as a post-Fordist regime of accumulation, the new economy or knowledge economy, the emergent geography of this new order is marked by the rise of regions as the salient foci of wealth creation (Martin and Sunley 2003). Characteristic in their high rates of technological learning and capacities for innovation, the workings of these regional industrial complexes, regional innovation systems or clusters have become a fixation for policy-makers and academics.[1] In particular, scholars have sought answers to the following questions: Why do certain regional economies perform better than others? What are the concrete mechanisms through which firms' capacities for learning and innovation are enhanced as a function of their spatial proximity? How are clusters inserted into knowledge spillovers at different spatial scales? And what are the ongoing constraints on everyday activities of learning, innovation and growth amongst knowledge-intensive firms in industrial clusters? In so doing, their aim is to provide guidance to local and regional policy-makers keen to promote local economic competitiveness, beyond mere lists of the 'critical factors' deemed necessary for cluster development (Markusen 1996; Wolfe and Gertler 2004: 1072).

In grappling with these questions, regional scholars have drawn heavily on the pioneering work of Alfred Marshall (1890) on nineteenth-century industrial districts, published in his *Principles of Economics*. Here, Marshall theorised a simple 'triad of localisation externalities', as sources of economic advantage that reside outside of the boundaries of individual firms but on which proximate firms can draw to improve their efficiency, productivity and competitive advantage.[2] Scholars have argued that contemporary high-tech regional economies display many of the relational characteristics and driving mechanisms of the urban craft communities that Marshall wrote about. In particular, they have revisited, updated and refined Marshall's notion of a regional 'industrial atmosphere' (or 'secrets of industry ... in the air'), in which he stressed the inseparability of industry from local society, traditions, social norms and values that are critical for economic coordination (Scott 2000: 30).

Within this socio-economic framework, the learning advantages that accrue to firms through spatial proximity are now widely recognised to emerge from enhanced access to networks of knowledge spillovers *between* firms (customers, suppliers, competitors), research organisations (universities, training institutes) and public agencies (technology transfer agencies, development agencies) (see Figure 2.1). As cheaper and more extensive communications technologies have made formalised (or 'codified') knowledge more readily accessible, scholars have explored the variety of channels through which firms are able to access less ubiquitous forms of 'tacit knowledge' (Polanyi 1967), which are more personal and embodied (Nonaka and Takeuchi 1995). These channels include: networks

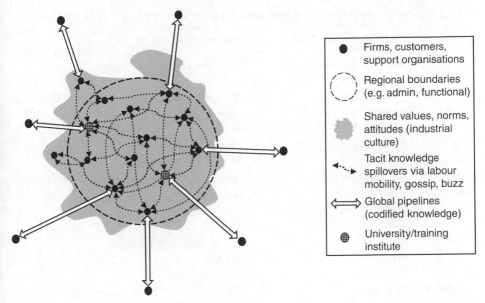

Firms, customers, support organisations

Regional boundaries (e.g. admin, functional)

Shared values, norms, attitudes (industrial culture)

Tacit knowledge spillovers via labour mobility, gossip, buzz

Global pipelines (codified knowledge)

University/training institute

Figure 2.1   Unpacking the regional learning and innovation agenda: key concepts/objects of study. (Modified from Bathelt et al. 2004.)

of repeated face-to-face interaction, peer observation, rumour and gossip (Henry and Pinch 2000); staff turnover and cross-firm career trajectories (Power and Lundmark 2004); and new firm formation and spin-off (Mayer 2013).

And beyond firms' abilities to *access* external sources of new knowledge, researchers have also analysed firms' abilities to reconfigure, transform and *use* that new knowledge to commercial ends; that is, their 'absorptive capacities' (Cohen and Levinthal 1990). Here, geographers have argued that shared sets of languages, norms and interpretive schemas enable economic actors to process imperfect information more effectively by reducing uncertainties when searching, screening, decoding and selecting information, and therefore to learn and innovate more effectively (e.g. Amin and Cohendet 2004; Malmberg and Maskell 2002). Accordingly, this work has also included a focus on corporate cultures as powerful influences on firms' capacities for learning and unlearning new behaviours (e.g. Schoenberger 1997; James 2005).[3] In short, it is difficult to explain the continuing competitive advantage of certain regional economies over others if their cultural conventions, relational networks and socially accepted norms of behavior are *not* taken into account (Storper 1997).

Importantly, these ideas concerning the learning and innovation advantages of 'being there' (Gertler 2003) have been developed through empirical engagement with clusters of knowledge-intensive firms in a range of high-tech industrial sectors, with a particular focus on North America and Western Europe, and more recently Southeast Asia and other sites in the Global South (see Table 2.1).

Table 2.1   Grounding regional learning and innovation theories: illustrating the scope of empirical research.

Country	Cluster example/s	Industrial sector/s
Australia	Melbourne, Sydney	Information technology, telecommunications, biotech, media
Brazil	Campinas, São Paulo	Computer software, telecommunications, microelectronics
Canada	Montreal, Toronto, Waterloo	Biotech, life sciences, ICT
China	Beijing	ICT
Cuba	Havana Science Pole	Biotechnology
Denmark	Aalborg	Wireless communications
France	Sophia Antipolis	Microelectronics, computer software
Germany	Baden Württemburg	Automotive
India	Bangalore ('India's Silicon Valley'); Bollywood, Mumbai	Computer software, IT; film production
Ireland	Dublin	ICT, computer software, finance
Israel	Silicon Wadi	ICT, computer software
Mexico	Guadalajara	Information technology
South Korea	Gumi City, Gyeonggi	Electronics, automobile components, semi-conductors, pharmaceuticals
Sweden	Kista Science Park (Stockholm), Medicon Valley (Oresund)	ICT, life sciences (biotech, pharmaceuticals)
Taiwan	Hsinchu-Taipei ('Silicon Island')	IT, hardware, semiconductors
Turkey	Ankara	Electronics, defense, software
UK	Motorsport Valley Oxford; Cambridge ('Silicon Fen'); City of London	Motor sport industry; information technology, computer software, electronics, biotechnology; financial services, law, business services
USA	Silicon Valley, CA, Orange County, CA, Seattle, WA, Portland, OR ('Silicon Forest'), Salt Lake City, Hollywood, CA	Semiconductors, computer software, microelectronics, internet, biomedical, aerospace, information technology, film and television

Unlearning Regional Learning? Three Analytical Blindspots

Unrelenting in their efforts to demystify the socio-economic foundations of regional advantage, successive cohorts of scholars have analysed in ever finer detail the multiple variants of regional industrial systems, and the variety of knowledges that circulate within and between them. For example, moving beyond earlier distinctions between tacit and codified knowledge, scholars have distinguished between analytical (science-based), symbolic (arts-based) and synthetic (engineering-based) knowledges (Martin and Moodysson 2013); between 'component' and 'architectural' knowledges (Pinch et al. 2003); and between 'know-what', 'know-why', 'know-how' and 'know-who' (Lundvall and Johnson 1994). Scholars have also questioned the neat scalar juxtaposition of tacit knowledge with localisation and codified knowledge with globalisation, to map instead the multi-scalar boundaries of innovation systems (Asheim et al. 2011) and the 'deterritorialisation of closeness' (Bunnell and Coe 2001).

In so doing, these successive refinements have yielded an 'abundance of theoretical concepts and explanations' (Malmberg and Maskell 2002: 429). No doubt about it. Problematically, however, through this 'focus on the minutiae of change … the linkages between firms in economic clusters and the incredibly detailed mapping of learning processes … these studies are very partial' (Perrons 2001: 208). In this section I explore this ongoing partiality. I argue that despite widespread recognition of the socio-cultural and relational embeddedness of regional learning and innovation processes, there remains a series of peculiar biases and analytical blindspots in which: (i) workers' lived experiences of being used as 'human capital' inputs to knowledge production and exchange processes drop out of the analysis; (ii) the majority of regional learning and innovation analyses remain gender-blind, perpetuating a theoretical invisibility of female worker agency; and (iii) analyses fail to locate activities of knowledge production within wider networks of social reproduction and care, as if these activities are somehow ontologically separated off from the rest of life. These three limits to the regional learning and innovation literature – and their mutual connections – are explored below. I argue that these three blindspots have obscured a powerful set of gendered constraints on firms' capacities for learning and innovation in industrial clusters, rooted in stubborn conflicts between work, home and family. In combination, they motivate the alternative analysis developed in this book, which brings the regional learning agenda into new conversation with labour geography and feminist economic geography.

Where are the workers? Labour as a benign factor input to knowledge production

> Innovation, even technological innovation, has a distinctly human face. (Amabile and Conti 1997: 111)

Successive regional learning and innovation studies have repeatedly highlighted local pools of specialised labour and spillovers of human capital as central factors in explaining regional advantage (e.g. Florida 2002; Wolfe and Gertler 2004; Cooke et al. 2011). Yet despite this analytical prominence afforded to labour, 'the traditional starting point in the literature on innovation and technological change for most theories of innovation has been the firm' (Audretsch and Feldman 2004: 2714; Bathelt and Cohendet 2014). Accordingly, where 'labour' enters regional learning analyses, it invariably does so as a passive factor input to production, similar to capital, equipment or raw materials, whose quality (levels of education, access to training, wage costs, union membership) and spatial distribution are key to understanding firm behaviour and uneven patterns of regional development.

In short, people are reduced to 'variable capital, an aspect of capital itself' (Harvey 1982: 380–381).

This firm-centric view of labour is particularly apparent in the work of Michael Porter, whose cluster concept has been particularly influential. Parallel tendencies are also apparent in the geographical economics of Paul Krugman. Crucially, however, these reductionist conceptualisations of labour as human capital are not simply found amongst economists within the regional learning literature, but also amongst major contributions by economic geographers. Examples include the reduction of workers to 'knowledge externalities', 'human capital externalities' and 'human capital stocks and flows' (Boschma et al. 2013: 1615); and to 'competence portfolios' and 'carriers of knowledge' (Breschi and Lissoni 2001: 991). This predominant focus on the quality and skills of the labour force as factor inputs to learning and innovation processes is problematic because:

> Referring to workers' various attributes and skills in terms of 'human capital' simply reinforces labor's theoretical invisibility, for it treats workers not as social agents capable of making landscapes in their own right but, rather, as simply an aspect of capital. (Herod 2001: 22)

In short, within these analyses, 'it is not actual human beings who are an input into the production process, but one of their characteristics – their capacity to do work' (Block 1990: 75). Indeed, while some scholars have begun to call for greater research attention to be paid to 'the creation and renewal of skilled labour ... as a central part of the learning process within regions' (Asheim et al. 2011: 886), their concern remains to 'explore the role that human capital plays in shaping absorptive capacity and innovation performance' (p. 886), and on firms' 'strategic use of regionalized labor pools' (Christopherson and Clark 2007: 10). Accordingly, economic geographers have asked the crucial question: 'where are the *people*?' (Eriksson and Lindgren 2011).

In contrast to these conceptualisations of abstract labour as a 'modifying force' that is factored in to the decision-making of capital (Herod 1997), an alternative workerist lens of analysis is concerned instead to explore workers' everyday lived experiences of *being used* as inputs to knowledge production. Beginning in the 1990s, Andrew Herod has codified a critical 'labour geographies' research agenda characterised by 'an effort to see the making of the economic geography of capitalism through the eyes of labour' (Herod 1997: 30; see also Castree 2007; Lier 2007; Rutherford 2010). Within this framework, labour geographers have demonstrated empirically how workers are capable of actively making and remaking the geographies of capitalism and effecting positive changes in their work and employment conditions, rather than simply watching passively from the sidelines and being *affected by* the dynamics of economic change. They have also demonstrated how place matters in shaping patterns of 'constrained worker agency' (Coe and Lier 2011), focused on the 'specific spatial settings and contexts' in which 'specific employment practices, work cultures, and labour relations become established' (Martin 2000: 456; see also Peck 1996). However, this intellectual shift from a productionist 'geography *of* labour' to an alternative 'labour geographies' research agenda remains severely under-developed within the regional learning and innovation literature.

Some important exceptions include Chris Benner's (2002) analysis of high-tech workers' experiences of volatile 'flexible' labour markets in Silicon Valley, and on the role of public and private sector labour market intermediaries in brokering those worker experiences and differently promoting cross-firm career advancement (Benner et al. 2007). Regional learning research in Australia has also explored how networks of informal information sharing in industrial clusters are shaped by labour relations; and shown how the sharing of high-value knowledge with competitor firms can be understood as a form of worker resistance (Wilson and Spoehr 2010). But beyond these studies, the regional learning and innovation literature offers minimal engagement with the labour geographies research agenda. Consequently, we know very little about workers' everyday lived experiences of being used as labour inputs to firms' knowledge production processes; nor of the geographical possibilities for knowledge workers to negotiate work-life improvements within and across the same knowledge-intensive firms and high-tech clusters which have long formed our primary objects of study.

Gender silences, theoretical invisibility of female worker agency

Closely linked to the analytical limits of the geographies *of* labour approach identified above, a second major blindspot within the regional learning literature concerns the theoretical invisibility of female worker agency; that is, of the active, everyday role of female knowledge workers in (re)making regional geographies of learning and innovation in practice. These problems are reinforced by a larger silence 'on the issue of gender despite its significance in the structuring of the labour market and in the organisations that make up the networking arrangements upon which the [learning region] concept so heavily depends' (Rees 2000: 89; see also Horelli 1997).

In part, these exclusions might be understood as resulting from the male-dominated sectors on which regional learning scholars have focused their attention, which include computer software, semiconductors, the motorsport industry, and engineering (see Table 2.1). But more than this, these exclusions are also rooted in stubborn social constructions of high-tech industries (which have formed the core empirical focus of the regional learning literature) as *masculine*. Here, then, scholars have repeatedly documented the occupational masculinity of engineering and the IT sector as our objects of study, rooted in independence, individualism, and 'requiring aggressive displays of technical self-confidence and hands-on ability for success, defining professional competence in hegemonically masculine terms and devaluing the gender characteristics of women' (Wright 1996: 86).[4] Crucially, however, Massey has also argued that the origins of these masculinist cultures must be understood not so much in terms of overt discrimination or sexism around the perceived unsuitability of women for high-tech jobs, but *ultimately* in terms of a set of deeply internalised broader societal dualisms which structure workplace social relations based on the logical and scientific nature of high-tech work:

> These jobs represent an apex of the domination of reason and science. It is this which lends them much of their status ... They are in a sector of the economy whose prime characteristics are, for these employees, structured around one of the oldest dualisms in Western thought – that between reason and non-reason; it is a sector identified with that pole – reason – which has been socially constructed, and validated, as *masculine*. (Massey 1995: 489, emphasis added)

Similar arguments around hierarchical dualisms that connect science with masculinity are also articulated in the work of Sandra Harding (1991, 2004). Her feminist standpoint critique of the natural sciences seeks to expose the androcentric bias (and hence incomplete manner) in which scientific *problems* are defined, how scientific *knowledge* is produced, and how scientific *rationality* is applied through new technologies. At the core of her critique, Harding argues that 'the androcentric ideology of contemporary science posits as necessary, and/or as facts, a set of dualisms – culture vs. nature; rational mind vs. prerational body and irrational emotions and values; objectivity vs. subjectivity; public vs. private – and then links men and masculinity to the former and women and femininity to the latter in each dichotomy' (1986: 136). Building on these ideas, Blake and Hanson (2005) have subsequently argued that masculinist assumptions are built into the very definitions and sites of technological innovation that economic geographers hold dear: in other words, regional learning and innovation analyses typically assume a male innovator.

Further reinforcing the theoretical invisibility of female workers within the regional learning literature, feminist scholars have also identified the subtle ways in which the very definitions of 'economy' are also riven through with masculine hierarchical assumptions and metaphors that remain stubbornly secure (e.g. Nelson 1995; Ferber and Nelson 1993, 2003; McDowell 2014a, b). Here, then, economic analysis is blessed with 'a natural unit of measure – money' (Nelson 1992: 109), in which price is the only relevant form of communication and representation of value. Accordingly, 'as those activities traditionally associated with women are not valued, at least in monetary terms, so they are worthless and excluded from the definition of what is necessary or essential' (McDowell 2000: 500). In seeking to expose and challenge these biases, significant advances have been made by feminist economic scholars to explore, for example, how much childrearing and household work contribute to the 'wealth of nations' and to challenge the invisibility of women and families in analyses of economic behavior (e.g. England 2003; Berik et al. 2009; Benería 2012; van Staveren et al. 2012). However, while much 'universal' economic analysis has been 'brought down to earth and given a pair of pants' (Bordo 1990: 137), this is less than apparent within the regional learning and innovation literature.

Thus while regional learning scholars have placed strong emphasis on sets of *shared* norms, practices, attitudes and expectations amongst firms and managers ('industrial cultures') which facilitate and enable patterns of interfirm interaction and foster mutual interfirm learning (Gertler 2002: 112), their masculinist constitution means that these norms are far from fully 'shared'. This is problematic given the use of regional learning analyses to inform regional policy and cluster initiatives, in a manner that risks reproducing the exclusion of female workers in other regions (Rees 2000).

Knowledge production abstracted from social reproduction

Closely linked to the analytical priority afforded to firms, and a general analytical silence around gender within the regional learning literature, a third blindspot concerns the false abstraction of networks of knowledge production from wider networks of social reproduction. That is:

> to questions in the immediate sense of the social division of labour between different kinds of paid work and between paid work and caring, or the wider sense of establishing sustainable regional development ... these dimensions are central to understanding the well-being of people within regions and therefore to regional or spatial development as a whole. (Perrons 2001: 211)

Instead leading scholars have tended to 'abstract clusters from the rest of the [socio]economic landscape, so that they often appear as isolated and self-contained entities ... as if they are isolated islands in the economy' (Martin and Sunley 2003: 17–18). 'Regional worlds of production' are analytically divorced from 'regional worlds of social *re*production'. 'New industrial spaces' are examined in isolation from 'new *re*productive spaces'. And 'innovative milieux' are analysed in isolation from 'social *re*productive milieux'. This disembedding of networks of knowledge production and exchange from wider networks of social provisioning is problematic given that labour's 'capacity to do work is not innate; it is socially created and sustained' (Block 1990: 75) – typically with women bearing the brunt of these tasks. Thus whilst feminist geographical deconstructions of 'life's work' (Mitchell et al. 2004) have documented the vital economic role of hidden household caring activities in sustaining the formal economy – indeed, also transforming the very notion of 'the economic' itself (McDowell 2000; Perrons 2010) – there remains a general lack of engagement with these intellectual advances within the regional learning and innovation literature. This strange state of affairs is also surprising given recent concerns to expand the focus of earlier partial analyses of the structures and forces shaping 'regional' processes of innovation and learning that were based on a misplaced conception of regions as 'closed systems' or mere 'containers of intangible assets and structures' (Yeung 2005: 47).

The overall result is that regional learning analyses have largely failed to explore the household support structures and institutional arrangements that enable socially *sustainable* innovation and worker learning; as if the knowledge production activities in which workers engage on a daily basis 'can be separated analytically and ontologically from their wider existence' (Castree 2007: 859) – or what Paula England has referred to as 'the separative-self bias ... that glorifies men's autonomy outside the family' (1993: 49). In this way, the regional learning agenda perpetuates a wider dualism between the sphere of transcendence (science makes breakthroughs, is involved in change, in progress) versus the sphere of immanence (the static realm of living in the present, or simple reproduction). And 'again it is transcendence which has been identified and constructed as masculine (he who goes out and makes history) as against a feminine who "merely" lives and reproduces' (Massey 1995: 189–190). The intellectual roots of 'masculine transcendence' can be traced to the long shadow of neoclassical economics, where the central character of the rational, autonomous agent makes rational, self-interested, optimising choices: 'He has ... no dependence on anyone, no responsibility for anyone but himself ... Economic man interacts in society without being influenced by society' (Nelson 1995: 135). And while *homo economicus* has been widely criticised as an inadequate model for female workers, it is also an inadequate basis for conceptualising male workers (England 2003; see also Barnes 1987). Or as Cynthia Negrey neatly puts it, 'we're employed as individuals, but we exist in families' (2012: 96). A key question for regional learning and innovation scholars therefore remains: 'if we were to change the central character in our economic story from the radically autonomous, isolated agent ... to the socially and materially situated human being, what effect would this have?' (Nelson 1992: 118).

Blinkered Economic Geographies – How Did It Get To This?

While the regional learning and innovation agenda has yielded multiple intellectual advances over the last three decades, it remains problematic on the basis of its tendencies to reduce workers to factor inputs to production, its general silence around gender and female worker agency, and its abstraction of networks of regional learning and knowledge production from wider networks of social reproduction. The intellectual lineage of these blindspots and partial frameworks of analysis can be understood in relation to a number of highly influential 'seed' studies. Cruz and Teixeira (2009) document the evolution of this research agenda through a bibliometric analysis of 37,531 different articles or books, as cited in 1780 articles on clusters and regional learning, using *Regional Studies* as the seed journal (1962–2008). Casting the net more widely across multiple journals, Lazzeretti et al. (2014) identify the 'key founders of clusters research' from a set of 1586 articles on clusters and industrial districts published in 250 international journals (1989–2010) using the ISI Thomson Reuters Web of Science database. As the two most highly cited authors in both studies, these analyses confirm the powerful influence of Michael Porter's (1985, 1998) work on clusters, and Paul Krugman's (1991, 1993) geographical economics analyses, with their focus on spatially uneven labour skills and wage costs as important 'factor input conditions'. They also evidence the strong international influence of work by Marshall (1890 [2013] and Scott (1988), whose emphases on firms' access to 'specialised labour pools' also sit squarely within the geography *of* labour tradition, rather than 'an effort to see the making of the economic geography of capitalism through the eyes of labour' (Herod 1997: 3).

Significantly in terms of understanding the origins of the masculine framing of the regional learning and innovation literature, and common tendency to divorce networks of knowledge production from networks of social reproduction, both bibliometric studies also point to the major influence of AnnaLee Saxenian's (1994) seminal study *Regional Advantage*. Indeed, Lazzeretti et al. (2014) also identify this as the single most important 'boundary spanner' seed study in their ISI database, based on frequencies of citation by scholars across different disciplines (economics, economic geography, business strategy, entrepreneurship and innovation studies), and different geographical areas (pp. 36–37).

Saxenian's (1994) study explored the divergent regional economic growth trajectories of California's Silicon Valley and Boston's Route 128 through the 1980s. Following sustained economic growth through the 1960s and 1970s, both regions faced economic downturns. While Silicon Valley rebounded with a new generation of start-ups, impressive technological performance and stabilised world market shares in semiconductors, Route 128 floundered. Shifting the analytical focus beyond a narrow concentration on transaction cost reductions, she argued that:

> Their differences in performance cannot be explained by approaches that view firms as separate from the social structures and institutions of a local economy. Variations in local institutions and corporate forms shape regional capacities for adaptation … the shared understandings and practices that unify a community and define everything from labor market behavior to attitudes towards risk taking. (1994: 6–7)

In Silicon Valley, a distinctive regional Californian counter-culture – characterised by an openness to experimentation, glorification of risk-taking, acceptance of failure as a learning process, rapid change as the norm and greater loyalties to transcendent technologies than to individual employers – is argued to have underpinned a regional, decentralised, network-based industrial system of learning. In contrast, a traditional conservative East Coast business culture in Route 128 – characterised by strong civic ties, self-sufficiency, risk aversion, stability, corporate loyalty and a stigmatisation of failure – is argued to have sustained relatively integrated corporations, a strict separation of work and social life, 9–5 work hours, lower levels of interfirm networking and labour mobility, an isolation of producers from external sources of know-how, and hence lower rates of learning, innovation and growth.

Central to her explanatory account is the heroic figure of the young, carefree Silicon Valley 'pioneer engineer' at the frontier of technical knowledge production (p. 30). Typically in their 20s and 30s, many of these men had migrated to California from the East Coast. Resulting from a lack of local family ties, many of these young, free and single engineers are argued to have developed a shared identity which blurred the work–home boundary. This reinforced a regularity of 'extraordinarily long work hours' and the completion of large workloads in short periods of calendar time (as well as contributing to high rates of worker burnout in the region (p. 46)).

Multiple generations of graduate students, academics and policy analysts have since applied and extended Saxenian's ideas to many other regional case studies worldwide (this yielding 11,024 citations and counting!).[5] Yet women are given virtually no labour agency within Saxenian's study. Indeed, only 15 of her 143 research participants were women. Path dependent from her work, subsequent regional analyses have also continued to divorce regional networks of knowledge production from those of social reproduction – because for Saxenian's Silicon Cowboys, work *was* their life, neatly fulfilling the masculinist myth of the 'ideal worker' (Williams 2000), for whom work is primary, time available to work unlimited, and the demands of family and personal life insignificant.[6]

In seeking to engender a more socially inclusive regional learning agenda, it is crucial that we recognise Saxenian's (1994) account as specifically grounded in the work-life experiences of a cohort of male engineers typically working in small start-ups, and at a particular moment in the historical development of this industry. Similarly, she also documented these men's work-lives at particular points in their *own* lifecourses when many were young, single, without dependents and hence able to dedicate themselves fully to the project of advancing new technologies. Arguably, this recognition of the temporal and geographical specificities of her account remains somewhat under-developed in many of the subsequent regional learning and innovation analyses that collectively represent her intellectual legacy.

(Re)gendering Regional Learning and Innovation

Rethinking innovation so as to recognise … how gender operates to enable or constrain certain innovative activities holds important implications for research and policy. (Blake and Hanson 2005: 697)

Despite its expansive nature, the regional learning and innovation literature remains blinkered in its predominant firm-centric focus, its analytical silence on female worker agency in actively (re)producing regional geographies of learning, innovation and growth, and its false abstraction of networks of knowledge production and exchange from gendered networks of social reproduction and care. At the same time, many feminist geographers steer well clear of what they perceive to be an increasingly abstract, self-referential and myopic regional learning and innovation literature narrowly focused on the minutiae of firms' productive processes at the expense of broader societal concerns of gender inequality and female worker empowerment. (This latter view was particularly evident throughout the review process for this book: 'the economic geographers won't read it … just cut Chapter 2!') Encouragingly, however, a select group of scholars have recently begun to challenge this subdisciplinary divide, making links between these two expansive research agendas to explore how the socio-economic dynamics of high-tech regional development are intimately connected to gender divisions of labour, uneven patterns of workplace inclusion and labour market inequalities. And at the root of their analyses of female worker agency, geographers are also responding to the increasing absolute (and relative) numbers of women within the various industrial sectors and occupations that are conventionally grouped together under the banner of 'high technology' and their unevenness at different geographical scales (nation-state, region, industry, occupation and firm).

The USA is the nation whose high-tech clusters have been continually hailed as exemplars and sources of inspiration of what policy-makers elsewhere can do to promote socio-economic development in the knowledge economy. In sum, women now account for over 1 million workers in US high-tech occupations: over 240,000 female software developers, 188,000 female computer systems analysts, 125,000 female computer support specialists and 100,000 female computer programmers.[7] Indeed, women now account for one third of all web developers and computer systems analysts in the USA. Women-owned companies also account for over 173,000 US high-tech firms, including 72,000 firms specialising in computer systems design and related services.[8] Women also account for significant proportions of high-tech labourforces in the European context, also prime empirical terrain for the expansive regional learning and innovation research literature. Whilst also typically marginalised within those European studies, women now comprise an average 33% of high-tech workers in the European Union (EU), ranging from figures of 44% (Bulgaria), 31% (Ireland) and 28% (UK), down to 23% in the Netherlands (Eurostat/European Union 2015). Indeed, (slowly!) increasing female high-tech labourforce participation is also underpinned by growing institutional efforts to promote and support female graduates in SET disciplines, and to encourage women into SET industrial sectors (interesting examples include the Athena SWAN (Scientific Women's Academic Network) initiative in the UK, and in the USA, the Women In Technology Project).

Against this varied industrial, occupational and institutional backdrop, recent work has begun to explore the gendered labour geographies of regional learning. This nascent body of work is introduced below. Crucially, it highlights the need for regional scholars to do more than simply add in gender as an additional category of analysis that was previously 'tucked away in an odd corner and presumed to be of interest only to women' (McDowell 2000: 498). Fundamentally, it has begun to expose the masculinist bias of a series of core concepts at the heart of the regional learning agenda, and which figure prominently in so many undergraduate and graduate economic geography programmes worldwide. The aim of this book, then, is to extend that critical project through new engagements with work-life balance (see Chapter 3).

Exposing the masculinist basis of 'buzz' and high-tech 'communities of practice'

In an effort to understand the cross-firm knowledge spillovers which develop within industrial clusters, scholars have articulated the concept of 'local buzz' (e.g. Storper and Venables 2002). In short, co-presence is argued to generate multiple planned and chance opportunities for personal meetings and communication, and for the development of cross-firm 'communities of practice' (Brown and Duguid 2000; Lave and Wenger 1991) whose members share similar technological knowledge, work routines and accumulated experience with particular problem-solving techniques. As employees swap knowledge and ideas about how things are done in other firms, ideas become recombined in new ways with firms' existing skills, technology, know-how and experience, hence stimulating innovation (Capello 1999; Lawson and Lorenz 1999). As shown by Saxenian's research in Silicon Valley, employees are understood to meet frequently not only at trade shows, industry conferences, seminars, talks and other social activities organised by local business organisations, but also informally in venues such as bars, clubs and cafés. In these social contexts, relationships are easily formed and maintained, technical and market information exchanged, contacts established and new ideas conceived.

Recent critiques of the buzz concept have, however, identified a series of gendered constraints on many female workers' abilities to benefit from the much celebrated 'thick web of information, knowledge and inspiration which circulate between the actors of a cluster' (Bathelt 2008: 86), and hence to participate in the development of (masculinist) cross-firm communities of practice. In an interview-based study of 88 ICT workers in Cambridge's high-tech regional economy, Gray and James (2007) documented the multiple ways in which female knowledge workers are repeatedly excluded from key events and activities in regional industrial systems through which buzz is widely theorised to occur in practice.

In contrast to the blurred work and social identities identified amongst Saxenian's Silicon Cowboys, Gray and James's (2007) analysis evidenced patterns of social interaction among female high-tech employees instead characterised by 'a more rigid separation of work and social life, premised in turn on childcare and other family commitments, which these women bear the brunt of within the home' (p. 422). Working mothers in their study consistently outlined how they have been forced to adopt compromised levels of informal social networking relative to their male partners. Indeed, many of the male participants in Gray and James's (2007) study made clear that they are only able to attend these after-work functions because they are supported by female partners who do not. Moreover, for female research participants with *and without* children alike, extra-curricular social events that predominate within the Cambridge region were widely perceived as 'male bonding sessions', rooted in a shared masculine identity which they found hard to break (p. 424). Female engineers therefore tended to forego attendance at key events where they might otherwise reproduce and enhance the value of their own labour power (see also Shih 2006; Gray et al. 2007; Durbin 2011). These exclusions also constrain female ICT workers' abilities to act as agents of informal information exchange between firms. Similar masculinist exclusions around networking have also been identified by female high-tech workers in Silicon Valley (Melymuka 2000; Hui 2014) and in Ireland's IT sector (Cross and Linehan 2006).

And in terms of intra-firm communities of practice, Gray and James (2007) also identified significant gender inequalities in the abilities of female employees to make their voices and ideas heard relative to their male colleagues. Specifically, female research participants outlined how they had articulated important insights and ideas that were either not incorporated into final product designs, or else were subsequently accredited to male colleagues. Similar experiences have also been documented through interviews with female engineers in Silicon Valley 'of working with men who … asked questions of male colleagues even though they knew I was most qualified to answer, or who seemed to resent the fact that I might be capable of coming up with better technical solutions on occasion' (Matthews 2003: 214).

Building female regional learning networks

In response to documented patterns of female worker exclusion in high-tech clusters, a second stream of work has explored the role of female-dedicated 'labour market intermediaries' in building female cross-firm occupational communities of knowledge exchange, upgrading skills and promoting female career advancement. At the forefront of this work, Chris Benner (2003) has documented the online and in-person networking activities of the Silicon Valley Chapter of Webgrrls (SVW) in (re)making regional geographies of learning and innovation.

Targeting female technology workers in a range of roles, Webgrrls International was founded in New York in 1995, subsequently growing to a membership of 30,000 women spread across over 100 chapters worldwide (Webgrrls 2007). Webgrrls enables female technology workers to 'stay on top of industry trends and changing skills demands, to find access to multiple employment opportunities when needed, and to build career mobility over time across multiple organizational contexts' (Benner 2003: 1819). To this end, Webgrrls combines online communications with monthly networking meetings in which female engineers dominate, and men are in the minority (they can attend only as the guest of a female member).

Other examples of female-dedicated LMIs in this genre are given in Table 2.2, and include Women In Technology, Girl Geek Dinners, Systers and the more recently formed Stemettes. While there exists some variation in the establishment, operation and points of intervention in the female career lifecourse amongst these different female-dedicated LMIs, their collective significance stems from their concern to increase rates of female labour market entry and subsequent career advancement within high-tech fields; and to foster female communities of practice within and across high-tech firms, educational institutions and funders. In so doing, they provide a highly visible and accessible set of infrastructures that currently enables thousands of female technologists (and aspirant labour market entrants) worldwide to share information, learn from credible role models, update their skills, raise their aspirations and build cross-firm careers. In short, these organisations enable female worker agency and (re)shape regional cross-firm dynamics of learning and information exchange in important – yet still under-researched – ways.

Towards a Socially Inclusive Regional Learning and Innovation Agenda

Over the last two decades, economic geographers have analysed in ever finer detail how the spatial co-location of firms in regional industrial agglomerations helps foster conditions conducive to learning, innovation and economic competitiveness. This literature has yielded important insights into the socio-cultural foundations of regional economic advantage; fundamentally shaped the content and focus of regional economic development policies worldwide; and become a major cornerstone of economic geography as a sub-discipline. Yet despite the expansive nature of this heterodox research literature and its international influence, the majority of its analyses remain conspicuously gender-blind. In response, the nascent body of research explored above has begun to examine the 'gendered geographies of high tech regional economies' (James 2008), and to give agency and voice to a set of workers generally marginalised in the regional learning literature. Rooted in a concern 'to start with an understanding of the variety and the reality of women's lives' (Rees 2000: 86), this work has documented

Table 2.2 Building cross-firm female communities of practice in high technology.

Organisation	Founded	Membership	URL	Description/Aim
Association for Women in Computing	1978	7 US chapters (plus overseas members): female computer programmers, system analysts, operators, technical writers, internet specialists, trainers and consultants	awc-hq.org	Professional development of women in computing through networking and programs on technical and career-oriented topics
Girl Geek Dinners	2005	Worldwide, women in the IT sector (industry, schools, universities): UK founded, over 1000 events to date in over 100 cities and 38 countries	girlgeekdinners.com	To inspire a new generation of female technologists and support those in the industry at present
MentorNet	1997	USA, women in engineering and related sciences: over 32,000 mentees paired with mentors in its first 15 years of existence; new membership alliance with LinkedIn	mentornet.org	US online mentoring network: connects senior women in technology to female and minority university students pursuing science, technology, engineering and maths careers
Stemettes	2014	UK membership (students aspiring to STEM profession, women already working in STEM, and STEM companies)	stemettes.org	To help combat the lack of women working in STEM and to encourage more girls to become creators and not passive users of new technologies. Connects female STEM students in UK schools with female STEM professionals (hands-on workshops, inspirational talks, personalised mentoring and funding opportunities)
Systers	1987	Over 4000 members across 54 countries worldwide: women technologists of all ages and at any stage of their studies or careers	anitaborg.org/get-involved/systers	Global online community of technical women in computing and technology fields: provides a private space for women to seek advice from their peers, and discuss the challenges they share as women technologists

(Continued)

Table 2.2 (Continued)

Organisation	Founded	Membership	URL	Description/Aim
Webgrrls International	1995	50 chapters across the USA, plus 12 international chapters (e.g. Denmark, Austria, Malaysia, Ireland, Canada, Philippines, Australia, Japan, New Zealand)	webgrrls.com	To encourage women to learn about, embrace and leverage the technology and the tools of the internet, to help them propel their careers and businesses forward and help them strive for and achieve success
WES (Women's Engineering Society)	1919	17 regional clusters in the UK – members from science and technology industries	wes.org.uk	Professional network of women engineers, scientists and technologists offering inspiration, support and professional development. Campaigns to encourage women to participate and achieve as engineers, scientists and technical leaders
WISE	1984	UK national membership includes individuals in industry and teaching (target of 1 million more women into UK STEM workforce), also MNC members	wisecampaign.org.uk	Promoting female talent in science, technology, engineering and mathematics (STEM) from classroom to boardroom – to increase the gender balance in UK's STEM workforce (30% female by 2020)
WITS Ireland	1990	500 members (2017) Ireland: ranging in age and experience from university students to some of Ireland's most eminent senior academics and business people	witsireland.com	An active forum supporting women in science, technology, engineering and mathematics (STEM) to reach their full potential through informed choices
Women In Technology	2004	UK female technologists, 7000 members (2011) at all levels: from graduates interested in a technology career, to senior managers and female returners from a career break	womenintechnology.co.uk	An online job board, recruitment and networking forum for women working in the technology profession in the UK. Seeks to help women succeed in an increasingly technical workplace and world through networking, exchanging job and business leads, and teaching new skills.
WITI (Women In Technology International)	1989	2 million members USA and worldwide, including Hong Kong, UK, Australia and Mexico	www.witi.com	To help women advance by providing access to – and support from – other professional women working in all sectors of technology

multiple exclusions of female engineers from networks of informal information sharing and buzz, masculinist communities of practice within high-tech firms and the development of female-dedicated cross-firm learning infrastructures. In short, by 'moving outside what are still considered to be the "normal" territories of economic inquiry' in regional learning and innovation analysis, this work has demonstrated how 'a whole new world hoves into view' (Thrift and Olds 1996: 311).

Problematically, however, beyond this small subset of studies, the mainstream regional learning and innovation agenda continues to maintain a general silence on the role of female worker agency in actively (re)producing regional geographies of learning, innovation and growth. It also continues to abstract networks of knowledge production and exchange from extra-firm networks of social reproduction and care. In this way, the regional learning agenda has remained strangely insulated from simultaneous debates within feminist economic geography (and beyond) focused on the shifting boundaries between work, home and family that have accompanied the transition to the knowledge economy, and the stubborn resilience of gender inequalities in paid work and caring (see, e.g., Perrons 2003; McDowell 2004; Mitchell et al. 2004; England and Lawson 2005; Perrons et al. 2006; McDowell and Dyson 2011). The overall result is that 'the wider consequences of economic change or firm competitiveness for the well-being of people in places are correspondingly neglected … [which] leads to analyses of spatial clusters or *development within regions* rather than the development of the region as a whole' (Perrons 2001: 208–210).

A useful visual metaphor on the need to expand the narrow analytical focus of the regional learning and innovation literature in this way is shown in Figure 2.2, based on a well-known cartoon by Alain (1953). Following Haggett (1995), we might then interpret the evolution of this influential research literature, and its evermore detailed mapping of intra-firm and cross-firm learning and innovation dynamics, as 'a succession of research projects that are clearly articulated with one another, like the letters on the side of the boat. Since the internal logic and precedent are well established, there is little questioning of the whole enterprise of which the individual projects form part' (Haggett 1995: 143).

Building on these critiques – and informed by wider calls for a greater level of engaged pluralism across the multiple sub-disciplinary strands of human geography (Barnes and Sheppard 2010) – *Work-Life Advantage* brings the expansive regional learning and innovation literature into new productive conversation with the equally expansive research literature on the gendered labour geographies of work-life balance (WLB), social reproduction and care (see Chapter 3). Moving beyond a focus on 'labour' or 'human capital' as factor inputs to knowledge production, the book draws on key advances in labour geography and feminist geography to explore workers' everyday experiences of being *used* as labour, and how gendered identities, varied responsibilities of care and personal life interests beyond the workplace shape workers' (non-)participation in the relational

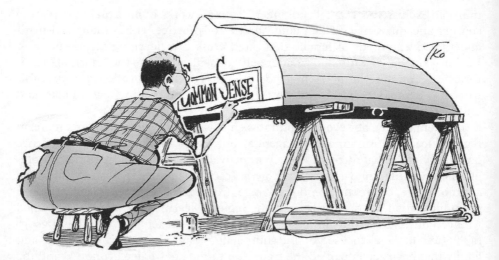

Figure 2.2   Engendering the regional learning agenda – a visual metaphor. Illustration by Tim O'Brien, inspired by a well-known cartoon by Alain (1953) originally published in *The New Yorker* magazine.

networks and communities of practice widely theorised as enabling regional learning and innovation. The aim, then, is to bring gendered work-life conflict, social reproduction and care to the core of an expanded regional learning agenda, rather than to sideline them as if ontologically separate from the 'real business' of knowledge production.

In this way, *Work-Life Advantage* is situated within a broader holistic economic development agenda (Pike et al. 2006), concerned to analyse the socially inclusive and sustainable nature of growth. Whilst certainly not excluding traditional economic concerns of competitiveness, growth and productivity, this research agenda attempts to move beyond economistic or 'dessicated indicators' (Morgan 2004) of growth and change. Instead it is concerned to articulate a wider and more rounded conception of high-tech 'development' that also integrates norma-tive questions around quality of life, gender equality and social well-being (Sen 1999; Rees 2000; Perrons 2004; Blake and Hanson 2005; Bristow 2005; Perrons and Dunford 2013). Accordingly, a holistic regional learning and innovation agenda is concerned to bring workers to the fore, to explore social inequities in the work-life experiences of similarly qualified workers in *doing* regional learning and innovation in practice. It is concerned to make visible the gendered networks of social reproduction and household divisions of labour which unavoidably shape workers' differential abilities and willingness to act as 'human capital' factor inputs to firms' knowledge production and exchange processes on a daily basis. It is concerned to explore how far the everyday workplace practices which collectively underpin the learning and innovation mechanisms widely celebrated in the regional learning literature might also be socially damaging, with negative implications for workers' health, well-being and quality of life across the work–home boundary. It is concerned to identify different types of working arrangements and employer interventions which might meaningfully reduce these negative outcomes for workers and their families; and in a manner that also enhances firms' learning and innovative capacities and long-term sustainable competitive advantage. And it is concerned to explore the spatial variability of high-tech development outcomes for workers and their families, within and between different regional economies and national welfare regimes. These concerns form the core focus of this book. In this way, my aim is to 'reright the boat' that currently constrains the wider impact, significance and social relevance of the 'mainstream' (read masculinist) regional learning and innovation research agenda.

Notes

1 Innovation may involve new product development based on research and development, or new process development based on the application of new technologies for continuous incremental improvements in the production process (Gray and Parker 1998). Either way, firms that innovate more consistently and rapidly typically employ more workers, demand higher skills, pay higher wages and offer more stable prospects for their workforce (OECD 2007b).
2 Marshall's triad of externalities comprise: (i) a local pool of specialised labour; (ii) local supporting and ancillary trades; and (iii) a local division of labour between firms. By many producers sharing the fixed costs of and access to common factors of production – land, labour, energy, transportation and other infrastructures – Marshall argued that the supply of such resources is enhanced as capital and labour migrate to these areas to take advantage of the larger combined markets for their services.
3 For useful introductory overviews of the expansive regional learning and innovation agenda see MacKinnon et al. (2002), Wolfe and Gertler (2004), and Boschma and Kloosterman (2005).
4 Other key studies that have differently analysed occupational masculinities in high technology include Wajcman (1991), Faulkner (2000), Panteli et al. (2001), Matthews (2003), Wilson (2003), Crump et al. (2007) and Peterson (2007).
5 Google Scholar identifies an impressive 11,024 citations for Saxenian's landmark *Regional Advantage* (1994) (as at 3 February 2017).
6 Saxenian also notes the role of long work hours in contributing to high rates of divorce amongst married engineers in Silicon Valley (1994: 46).
7 These figures as at May 2015. Source: US Bureau of Labor Statistics, Current Population Survey, www.bls.gov/cps/cpsaat11.htm.
8 Figures refer to privately held companies in which women hold a 51% or greater share of the company. Source: US Census Bureau, Survey of Business Owners 2012, https://www.census.gov/econ/sbo/.

Chapter Three
Work-Life Balance and its Uncertain 'Business Case'

Introduction

> Research into the ways families and the nature of work are changing has prompted legislative and workplace policies aiming to make it easier for people – particularly women – to combine paid work with the rest of life. Yet their impact is limited ... it seems societies are more or less stuck about how to go forward with equitable and sustainable change. (Lewis et al. 2003: 824–825)

> I think about work-life balance a lot. I have to. If my life were a household budget I'd be down to zero every month at best, overdrawn every week at worst ... It can often make you feel like an egg-timer on its last few seconds of sand. (Hobsbawm 2009: 2)

Work-Life Advantage brings the regional learning and innovation agenda into new productive conversation with major debates in labour geography and feminist geography, around the shifting spatial and temporal boundaries between work, home and family that have accompanied the transition to the 'new economy',[1] and the stubborn persistence of gender inequalities in paid work and caring. In contrast to celebratory policy rhetoric around new worlds of work, the experience of 'flexibility' for many workers has meant increased workloads, less predictable schedules and more unsocial work hours (Bunting 2005; Gambles et al. 2006; Lewis et al. 2007). At the same time, household life has also become more complex through increased female labourforce participation and increased numbers of dual-earner households. Simultaneously, the neoliberal rollback of social provisioning has transferred the burden of care down to the 'natural' level of

Work-Life Advantage: Sustaining Regional Learning and Innovation, First Edition. Al James.

home (Bakker and Gill 2003) where most women continue to bear primary responsibility for direct child-facing care and domestic tasks. The overall result is a complex, gendered and often painful set of conflicts between competing demands of paid work and personal responsibilities and life interests beyond the workplace. In response, the desirability and means of achieving an appropriate work-life balance (WLB) has received widespread attention from governments, employers, labour unions, academics and the media.

The societal significance of WLB is profound, with multiple studies documenting causal connections between poor WLB (termed work-life conflict) and heightened stress, reduced psychological well-being and quality of life, deteriorating familial relationships and ongoing gendered labour market inequality. In response, commentators have articulated repeated calls for enhanced WLB policy provision, framed in terms of moral obligation, social justice and equality of opportunity (e.g. Connell 2005; Lewis and Haas 2005). Crucially, however, scholars have also come to the reluctant conclusion that employers are unlikely to introduce work-life provisions for their employees unless they can identify bottom-line economic advantages from doing so. This so-called WLB 'business case' lies at the heart of government policy interventions in liberal welfare states, and is epitomised in the UK, Ireland and the USA. Here, employer benefits from work-life provision are widely promoted as improved recruitment, retention, morale and productivity, and reduced stress, absenteeism and costs. Yet despite its popularity, there remains a relative dearth of empirical evidence to support these claims in practice, perpetuating scepticism amongst many employers and ongoing hardship for many workers and their families. Exacerbating these problems, the global economic downturn created new work-life demands. And with employers keen to effect cost savings in pursuit of post-recessionary survival and growth, employer-provided work-life arrangements have been far from immune (Galinksy and Bond 2009).

This chapter offers a summary introduction to the expansive work-life research literature, as the second major research stream which frames this book. It begins by documenting the phenomenal rise of the work-life agenda over the last two decades in relation to a series of interrelated tipping points, and a broad range of studies that have repeatedly evidenced the labour market, health, family and community consequences of work-life conflict. This discussion also situates the rise of WLB as a set of practices and (contested) discourses rooted in a broader neoliberal political project designed 'to re-establish the conditions for capital accumulation' (Harvey 2005: 19). An analytical distinction is also made between work-life arrangements that *accept* the accommodation of working time to the gendered division of labour and those that also *challenge* the male carer model at a deeper level by supporting the reallocation of reproductive tasks within the household.

The second half of the chapter explores the increasing salience, policy appeal and multiple limits to the WLB business case. I argue that the impact of previous

WLB research has been limited by three major conceptual/methodological problems: (i) prioritisation of employer needs at the expense of workers and their families; (ii) limited focus on output measures of firm performance rather than the underlying sources of firms' long-term sustainable competitive advantage; and (iii) atomisation of firms from the regional industrial systems of which they form part. These three critiques are developed with reference to a range of influential WLB business case analyses. They also frame the alternative regional learning/WLB mutual gains analysis developed in Chapters 5, 6 and 7 concerned to identify the kinds of work-life arrangements that workers and their families find most useful in reducing work-life conflicts, to identify the concrete mechanisms through which different bundles of employer-provided WLB arrangements can also enhance firms' learning and innovative capacities, and to explore how those WLB-learning outcomes vary geographically within and between regional economies. The final section highlights the increased salience of this work-life/learning analysis in the wake of uneven geographies of economic crisis and recovery.

Rise of the Work-Life Balance Agenda: Why Now?

> Being against work-life balance would be a bit like being against summer or good sex. (Benn 2002: 18)

The work-life balance research agenda emerges from growing concerns around the invasiveness of paid work in people's lives, and the difficulties of successfully managing a job alongside care for dependents, friendships, communities, personal life interests and leisure (Lewis et al. 2003; Bunting 2005; Weeks 2011).[2] At its core is an explicit rejection of the glorified norm of the masculine 'ideal worker', for whom work is primary, time to spend at work unlimited, and the demands of family, community and personal life secondary. And much more than simply enabling women to work on the same terms traditionally available to men, the work-life research agenda challenges the very definition of the 'ideal worker' to reflect norms of parental care (Williams 2000), to think creatively about how to redistribute paid work and care through non-traditional working arrangements that change the conditions under which both men and women work, and to humanise the workplace for all workers who wish to lead rounded lives (Doherty 2004: 433).

Concerns around the social sustainability of work are certainly nothing new, with dilemmas related to the management of paid work alongside family life having been a focus of Anglo-American research from the 1960s onwards (e.g. Rapoport and Rapoport 1965; Kanter 1977).[3] This work emerged in response to the increased labourforce participation of women in the years following the Second World War, and the increased physical and mental strain exhibited by many workers based on limited job autonomy and a lack of employer support for quality of life (O'Toole 1974).[4] Extending this focus, subsequent research concerned with work-family balance continued through the 1980s, with its particular focus on working mothers as the cohort of workers with the most visible non-work demands (either in their support of a male breadwinner or as part of a dual-career couple). But in marked contrast to the scale of this earlier research, from the late 1990s onwards we have witnessed an 'explosion of research, media coverage, and political discussion … around issues of work and family' (Drago and Hyatt 2003: 139). In short, the work-life balance agenda has brought a new urgency to long-standing debates around the intimate (dis)connections between production and social reproduction, with WLB hailed as a 'hot topic' (Lewis et al. 2003), a 'powerful, new national sentiment' (Greenblatt 2002: 177) and 'the challenge of our time' (Gambles et al. 2006).

Evidencing this rise to prominence, WLB has commanded attention from governments at the national and international scales, including the US Department of Labor WLB Taskforce and White House Flexibility Forum and the UK's Family-Friendly Working Hours Taskforce; and, from 1998, has been more firmly integrated into the guidelines that accompany the European Employment Strategy (Lewis 2009). WLB has become a major campaigning focus for multiple advocacy groups pushing to secure a better deal for workers and their families, including, in the UK, the Work Foundation, Working Families, the Fatherhood Institute and Trades Union Congress; and, in Ireland, the Equality Authority and Irish Congress of Trade Unions. WLB has also become a common part of everyday discourse in employing organisations, with a proliferation of workplace training programmes on managing work, home and family, specialist WLB consultants and coaches, and countless websites offering advice and resources on how WLB can be achieved. Accordingly, companies jockey for position on the Working Mother list of the Best 100 Companies and Forbes Top 25 Companies for Work-Life Balance, alongside the growth of WLB-orientated employer organisations, including Employers for Work-Life Balance (UK) and Corporate Voices for Working Families (USA). National newspapers frequently run stories on the negative impact of overwork on the health and well-being of workers and their families, with WLB featuring on the covers of *Business Week*, *Harvard Business Review* and *Time Magazine*. It has also been identified as a pervasive management practice in Scott Adams's award-winning comic strip Dilbert no fewer than six times to date.[5]

At the same time, multiple research councils and philanthropic organisations have funded major research programmes on WLB and family-friendly employment, including in the UK the ESRC's Future of Work Programme and ESRC Gender Network, and in the USA the Families and Work Institute (funded by the Ford Foundation) and Sloan Research Centers for Working Families at multiple universities. Successive WLB surveys have repeatedly documented the scale of struggles experienced by workers and their families in

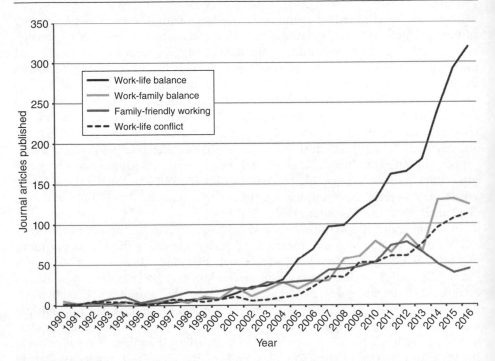

Figure 3.1   Documenting the exponential growth in WLB research. Source: Author's bibliometric search of Web of Science (accessed 7 March 2017).

the UK since 2000, alongside the inclusion of WLB variables on larger surveys including the UK's Workplace Employees Relations Survey, British Social Attitudes Survey, European Working Conditions Survey and the OECD Better Life Index. And all this alongside multiple international academic conferences on WLB, work-life themed journal issues and an exponential growth in the publication of academic journal articles on this topic from the late 1990s onwards (Figure 3.1). A similar pattern of exponential growth in academic interest in work-life balance can also be identified using Google Scholar, with a whopping 36,100 hits for 2004–2014 (cf. 16,400 hits for 1994–2004, and 3,720 hits for 1984–1994).

A number of interrelated drivers and tipping points (Kossek et al. 2011) have underpinned this historical ascendancy of the work-life balance agenda, understood both as a set of practices that respond to a negative set of changes experienced by workers and their families and as a WLB discourse perpetuated by policy-makers, employers and the media. The following sections explore four commonly identified sets of drivers, and the ways in which they powerfully shape the everyday conditions under which workers live, love and labour.

Changes in paid work: flexibilised labour, work intensification, time squeeze

Widely identified as a major driver of contemporary WLB concerns, the emergence of post-Fordist flexible production from the late 1970s has yielded dramatic changes in the organisation of paid work and working times. In response to heightened global competition, narrowing profit margins, impatient shareholder demands and the opportunities afforded by new technologies, employers have pursued a range of flexibilities in order to remain competitive. Numerical flexibility refers to firms paring back their regularly employed core workforce, who are supported by a less secure, peripheral workforce whose working times (and hence wage costs) vary with fluctuations in demand (Peck 1996).[6] Indeed, in the context of weakened employment laws, numerical flexibility is also achieved through hiring and firing apparently secure core workers (Burchell et al. 2002). Scholars also point to far-reaching changes associated with functional flexibility and multi-skilling, in which traditional job demarcations have been eroded, job contents redefined, and peer and authority relations transformed such that workers take on a broader range of tasks. The overall benefits to firms are hailed as increased responsiveness to volatile consumer demand, improved productivity, reduced operating costs and an apparent sense of empowerment amongst employees as they are allowed to control their work within multi-skilled teams rather in bureaucratic hierarchies.

However, whilst the benefits of numerical and functional flexibility are widely celebrated at the level of employers, workers' experiences of *being used* flexibly in this manner are often less than positive. In short, as firms have sought to displace the risks of market volatility, flexibility for many workers has come to mean working to less predictable and more individualised work schedules, and more unsocial work hours. In contrast to earlier realities of a fixed standard 9–5 work day, '24/7' has become common parlance to denote round-the-clock availability, resulting in increased stress around scheduling social reproductive tasks around paid work (Presser 2006). Indeed, the erosion of the collective rhythms of working life is also recognised as cumulative, because 'as working hours have become more varied, people will expect services to be available at a wider range of times' (Perrons 2003: 69). At the same time, the rise of 'high commitment workplaces' – in which core workers are also turned into their own managers – has often meant the inculcation of a work ethos that rewards being busy and looking busy (Rapoport et al. 2002). Thus for Bailyn (1997: 211), 'putting in time – being visibly at work, often for long hours – is often seen as a sign of commitment, of loyalty, of competence and high potential, and in many cases as an indicator, in and of itself, of productive output'. Here, commentators have highlighted the use of software programs designed to temporarily hold emails to be sent later in the evening; and employees leaving a coat in the office, or car in the car park, all to give the impression of being present at work longer (e.g. Perlow 1997).

Problematically, the pressure to demonstrate commitment by working long hours is also likely to be strongest in the early stages of the career, at the same time as many workers are starting families.

Closely linked to these temporal outcomes of workers being flexibilised, commentators have also highlighted pervasive problems of work intensification, understood as the increasing effort that employees put into their jobs during the time that they are working (Burchell et al. 2002; cf. Bonney 2005).[7] Marshalling evidence from an impressive range of datasets, studies have documented many workers now working at higher speeds and to tighter deadlines (Green and McIntosh 2001; Burchell et al. 2002).[8] Whilst positive assessments of work intensification might emphasise the ways in which professional jobs are more rewarding than in the past – hence that workers are simply more self-motivated (Florida 2002) – other analyses highlight problems of 'horizontal loading', in which employees are expected to take on extra tasks at the same (and indeed lower) skill level (Hudson 2002), increased workflow speed as new technologies structure the labour process more efficiently, and downsizing, which requires that remaining staff still achieve the same overall quantity of work. The pervasiveness of these practices is further reinforced by weakened employment laws, erosion of trade union power, tyrannies of performance management and employee perceptions of job insecurity that further motivate them to work harder (McGovern et al. 2007; Taylor 2013).

The outcome of these combined changes in paid work has been signified through the use of a range of metaphors including 'the time squeeze' (Schor 1991), 'the time bind' (Hochschild 1997), 'time famine' (Perlow 1999), 'time scarcity' (Negrey 2012) and 'overwork' (Bunting 2005), alongside a proliferation of statistics invoked to evidence the scale of this multifaceted problem. For example in the UK, a major study by the Joseph Rowntree Foundation found that 41% of working fathers started work early morning (6:30–8:30 am) several times a week, and 48% of working mothers regularly work at weekends (La Valle et al. 2002: 8). Bunting (2005: 18) has also calculated that British workers work almost eight weeks longer per year than do their European counterparts, as a function of longer work days and shorter holiday leave entitlement. More recently, the Trades Union Congress (TUC 2015b) calculated that the extent of unpaid overtime (1.2 billion hours) in the UK was worth £32 billion in 2014. And in a 2014 UK survey by international private healthcare provider Bupa, 28% of workers admitted to not taking a break of any kind during their working day[9] – consistent with regular news stories on the growing prevalence of 'al desko' dining (e.g. *The Guardian* 2014, 2015a), accompanied by frequent photo uploads documenting the everyday lived realities of this trend at saddesklunch.com. Similar trends have also been identified in Ireland through the National Workplace Survey (2003 and 2009), documenting a significant increase in work pressure over the last decade (Russell and McGinnity 2014).[10] Likewise, nearly three quarters of Irish workers eat lunch at their desks during their 22-minute lunch 'hour'.[11]

In combination, these results point to 'the increasing pervasiveness of paid work in people's lives' (Lewis et al. 2003). Significantly, research has also documented the rise of flexibilised labour, work intensification and high commitment workplaces in a range of high-tech sectors including information technology, semiconductors and biotechnology – that is, in the *same* industrial sectors that have long formed the objects of regional learning and innovation studies (see Chapter 2). However, the consequences of these common sources of work-life conflict amongst knowledge workers for everyday practices of learning and innovation remain largely unexplored.

Household change: working women and (post)modern family structures

A second major set of drivers underpinning the rise of the WLB agenda concerns dramatic changes in the structure of families and households over the last three decades, and the relative contributions of women and men to paid work. At the forefront of these shifts, household life has become more complex as female labour market participation rates continue to grow and an ever increasing proportion of workers become part of dual-earner households (Crompton et al. 2007; Esping-Andersen 2009). Epitomising these changes, the UK and Ireland have seen significant increases in female labourforce participation rates since 1990, and a reduction in the extent to which women are doing less paid work outside the home than men (Table 3.1). This upward trajectory in female employment has been particularly apparent amongst women aged 25–54 years old, the age range most likely to have dependent children (Table 3.2). Indeed, the majority of cohabiting households with children in the UK are now classified as dual earner (comprising 29% dual-earner full-time and 31% male full-time and

Table 3.1   Labourforce participation rates by sex: UK, Ireland, EU, OECD (1990, 2000, 2013).

	Men (aged 15–64 years) %			Women (aged 15–64 years) %		
	1990	2000	2013	1990	2000	2013
UK	88.3	84.1	83.1	67.3	68.9	71.7
Ireland	77.5	80.0	77.2	42.6	56.3	63.2
EU15	79.8	78.7	79.5	54.7	60.2	67.8
OECD	82.2	80.8	79.7	58.1	59.1	62.6

Source: OECD (2014) 'Labour market statistics: labour force statistics by sex and age: indicators', OECD Employment and Labour Market Statistics (database).

Table 3.2   Women's employment by age: UK, Ireland, EU, OECD (1990, 2000, 2013).

Women's employment to population ratios	15–24			25–54			55–64		
	1990	2000	2013	1990	2000	2013	1990	2000	2013
UK	65.9	59.1	48.7	68.6	73.1	75.3	36.7	41.4	53.0
Ireland	39.6	45.1	29.7	39.3	62.6	65.6	18.2	26.8	43.2
EU15	41.9	37.8	34.5	59.7	65.8	71.2	24.3	27.9	45.9
OECD	45.7	40.8	36.5	61.6	63.7	66.5	34.5	36.7	48.1

Source: OECD (2014) 'Labour market statistics: labour force statistics by sex and age: indicators', OECD Employment and Labour Market Statistics (database). This following Lewis (2009: 34).

female part-time – see Connolly et al. 2013), with this requiring the coordination of paid and unpaid work between partners who may be working different shifts (Lewis 2009). Similarly in Ireland by the later years of the Celtic Tiger growth period, dual-earnership had become more common (51%) than the traditional male breadwinner/female homemaker arrangement (36%) among working-age couples (McGinnity and Russell 2007).[12] And by 2015, the labourforce participation rate for women in couples with children was 65%, rising to 70% for those women in couples whose youngest child is aged 0–5 years (CSO, Irish Quarterly National Household Survey 2015).

As widely documented, this changing gender composition of the labour force is rooted in increased female educational achievement, alongside the dramatic growth of service sector employment, and its social construction as appropriately 'feminine' (Hochschild 1983; McDowell 1991). It also emerges from shifting societal constructions of the 'good mother' as she who enters the labour market to raise her income and skill levels for the benefit of her children, who no longer occupies the home as a continuous presence, and who hands over the care of her children to another for part of the day (McDowell et al. 2006: 145). Likewise, welfare state entitlement for women in their roles as mothers or wives has also become increasingly conditional on labour market participation (Perrons et al. 2006). Adding further complexity, increased rates of divorce and a growth in the number of women having children without marriage have also increased the numbers of female-headed single-parent households, alongside a shift towards greater 'family fluidity' (Lewis 2009), as people move in and out of marriage and cohabitation.

On one level, these interrelated shifts have presented an unparalleled challenge to the male breadwinner model previously dominant in the mid-twentieth century, which now accounts for only 22% of cohabiting households with children in the UK (Connolly et al. 2013) compared with 35.5% in Ireland (Lewis 2009).[13] However, there also remain stubborn gender inequalities in the distribution of childcare, an 'incomplete revolution' (Esping-Andersen 2009) in which women continue to assume the majority role for everyday tasks of unpaid housework and care, or the 'second shift' (Hochschild 1989). In other words, while women have increased their participation in paid work greatly, men have not increased their participation in unpaid household work to an equal degree, and this holds true for all developed countries (Gershuny 2003). And while research on working time shows that men in households with children typically work longer than men in general, when paid work and unpaid caring work are combined it is women who work longer hours on average and continue to experience the stress and anxieties of combining these roles (Moen 2003; McDowell et al. 2005; Lewis 2009). The perversity of this stubborn problem is neatly summed up by Cynthia Negrey, who points out that:

> For wives to make substantial contributions to paid work in response to household need, and yet for husbands not to make significant contributions to household labor time in response to such need, can be seen as utility maximizing only if wives, dead on their feet after a double day, are always more productive in household labor than their husbands. (2012: 29)

In addition, the double burden faced by many women is reinforced by the decline of the extended family; increased expectations of more intensive parenting (Bianchi 2000); and greater eldercare responsibilities in the context of increased life expectancy. Accordingly, these challenges have led to a greater appreciation of family (and its variety of forms) as an important independent variable by social policy analysts and policy-makers (Esping-Andersen 1999, 2009). In stark contrast, there remains a deafening silence around families, reproduction and care within the extant regional learning and innovation research literature (see Chapter 2).

Welfare change: neoliberal attack on social provisioning, commodification of care

A third set of drivers underpinning the rise of the WLB agenda concerns the restructuring of the welfare state as a major structural influence on how individuals and families organise their lives. Over the last three decades, as part of the neoliberal attack on social provisioning, the burden of care has been transferred down to households where most women continue to bear the prime responsibility for direct child-facing care and domestic tasks – or the 'messy and fleshy' components of domestic and family life (Katz 2001; Fawcett Society 2009).[14] At the same time, a shift from so-called passive to active welfare (workfare) also means that welfare entitlements are increasingly dependent on labour market participation. Thus while governments have encouraged women into the paid workforce, this has not been matched by adequate state-provided childcare to enable that transition. This is especially the case in the UK and Ireland, where the costs of private childcare are also prohibitive for many families. For example, childcare costs for a dual-earner family with two children are highest in the UK at 32.7% of net household income, followed by 29.2% in Ireland (OECD 2007a). Compare this with equivalent figures of just 6.2% in Sweden, or 4.7% in Belgium (all figures calculated assuming 2-year-old and 3-year-children, and after provider subsidies and tax benefits, with dual-earner couple's aggregate earnings at 167% of the country's average wage). In short, neoliberal welfare state cutbacks have therefore assumed an infinite capacity on the part of households to absorb the costs of care (MacDonald et al. 2005).

Motivating these changes, neoliberal claims for the apparent superiority of free-market principles and values of economic rationalism, competition, individualism and freedom have legitimised a set of welfare policy practices to increase economic productivity, competitiveness and growth, while reducing the responsibilities of employers and the state to sustain the social reproduction of labour-power (McDowell 2004; Mitchell et al. 2004). Thus for Staeheli (2013), the fundamental obligation of citizens to care is turned on its head in neoliberal policies centred on the individualisation of rights and responsibilities. Rather than care reflecting an obligation to one another to foster well-being and self-development, care instead becomes a responsibility to care for oneself, such that one does not place a burden on others (see also Lawson 2007). The result is a problematic search for personal, private solutions to problems of work-life conflict, rather than identifying with others and collectively campaigning for reform (Bunting 2005). These individualised 'solutions' range from the purchase of private childcare, to domestic cleaning services, to microwave meals, to rent a grandma services, and even outsourced ash scattering and gravesite maintenance – or what Hochschild (2003, 2012) has identified as the 'commercialisation of intimate life' and 'the outsourced self' as a means of buying in (depersonalised) care and social reproductive functions which we no longer have time to do (the market takes away with one hand but 'conveniently' gives back with the other). However, in the absence of adequate welfare provisions, certain 'biographic solutions to structural conditions' of work-life conflict are often only available to a minority of workers who can afford to purchase them through the market, in a manner that connects gendered work-life conflict to a perpetuation of class inequality.

Oddly, however, despite the major significance of these welfare changes and their implications for workers and families, welfare provisioning rarely enters into regional learning debates – as if these regional 'stocks of human capital' exist in strange isolation from its influences on how they live, love and labour.

Urban geographies of (dis)connection: housing, commuting, liveability

Alongside recent changes to the structure and organisation of paid work, families and households, and neoliberal welfare provision, a fourth set of drivers underpinning the ascendency of WLB debates stems from spatial differences in the urban 'spacing' of jobs, housing and public services which geographically delimit the work-life options available to workers and households. Studies have explored how the spatial fragmentation between housing, jobs, schools, transport systems, leisure and commodified forms of caring services (eldercare facilities, childcare services, gyms, hair salons) is crucial to understanding the wider context in which work-life calculations and decisions are made (England 1991) – or what Jarvis (2005: 141) identifies as the 'infrastructure of everyday life'. The coordination of everyday life within these urban 'carescapes' involves the delicate coordination of multiple tasks of production and reproduction of households undertaken by individual and collective actors, distributed across different (and often mismatched) spaces and times (e.g. McKie et al. 2002; Bowlby 2012). Here, Perrons et al. (2006) highlight expanding travel-to-work areas, longer and more intense rush hours and growing congestion associated with school runs as some of the major everyday challenges that face workers, families, employers and service providers in struggling to reconcile the daily complexities of social reproduction with waged work, around daily activities that are often not in close proximity to one another. These challenges are reinforced by suburban sprawl and the privatisation of public transport services in many cities, with an emphasis on profitable peak time routes (McDowell et al. 2006).

Strikingly, these urban geographies of disconnection have been identified not only in the same cities and regions that have also formed the core focus of the regional learning literature, but also as a direct *outcome* of innovation-led regional development in those places. At the forefront of this agenda, a report from the Silicon Valley Community Foundation (SVCF 2015) details how 65% of Silicon Valley CEOs identify employee housing costs as one of the top five challenges for doing business in the region, with Silicon Valley housing sale prices (proxied by San Jose) rising 33% between 2012 and 2014 alone (p. 36). In combination with high prices in Silicon Valley's rental housing market, families are forced to search out cheaper suburban housing locations further away from their place of work, with nearly 1 in 6 commuters travelling two hours or more each day in 2013, up from 1 in 8 in 2011 (p. 6). Additionally, Silicon Valley commuters are also identified as wasting 84 hours per year for a half-hour commute in 2013 due to delays from traffic congestion, second in the USA only to Southern California (proxied by Los Angeles) at 90 hours per year (p. 36). Further compounding these difficulties, recent commentaries have also highlighted the high costs of childcare as a 'dark side of Silicon Valley' (Forbes 2013) and limited quality of public transport provision (Bloomberg 2013).

Similar problems of urban work-life disconnection through reduced housing affordability, urban sprawl, childcare provision, public transport challenges and/or lengthy commutes have also been identified as a function of high-tech industrial growth in other well-known learning regions including Austin, Texas (McCann 2007), Seattle (Jarvis 2005) and Portland, Oregon, or Silicon Forest (Song and Knaap 2004). Encouragingly, therefore, in combination with the feminist geographical research outlined above, we might expect this work to have opened up new directions for the regional learning and innovation agenda. Not least given that these urban work-life challenges have been identified in some of the same high-tech regional 'blueprint' economies at the very heart of the regional learning and innovation agenda. However, this expanded conversation remains significantly under-developed in practice. In short, economic geographers need to get out more!

Work-Life Balance as a Contested Concept

While different commentators place differing explanatory emphases on these four sets of drivers, there is widespread agreement that their combined result is a complex, gendered and often painful set of conflicts between competing demands of paid work and personal responsibilities and life interests beyond the workplace, for which workers have only 'finite resources in terms of time and energy' (Cooper et al. 2001: 50). That is, a daily 'struggle to juggle' (Bacik and Drew 2006) between the availability demands of 'timelessness' required by employers and those of 'timeliness' required by the everyday events of intimate family and personal relationships (and most markedly the routine of children's lives) (Bunting 2005: 16; Lewis 2009). In response, the desirability and means of achieving an appropriate work-life balance (WLB) has received widespread attention from governments, managers, trade unions, academics and the media. At the individual level, the 'precarious accomplishment' (Perrons et al. 2006) of WLB refers to 'the absence of unacceptable levels of conflict between work and non-work demands' (Greenblatt 2002: 179), or 'the extent to which individuals are equally involved in – and equally satisfied with – their work role and family role' (Greenhaus and Singh 2003: 2). While encompassing earlier family-friendly perspectives, the linguistic shift to the WLB term was intended to broaden the debate beyond a narrow focus on working mothers, to engage with men and women with a range of responsibilities, commitments and personal life interests outside of paid work. Examples include people wanting time off or alternative working arrangements because of religious observances, continued education alongside paid work, sports, hobbies or community and charity work. Thus for Williams (2000), WLB is about balancing people's work-life needs to needs around care of self, and the maintenance of mind, body and soul with needs around caring properly for others and with people's needs to gain economic self-sufficiency. Significantly, the work-life agenda has also spread beyond its Anglo-American roots to other English-speaking countries (e.g. Australia, New Zealand, Canada) and to non-Western countries including India, Japan, China, Singapore and the Philippines. However, the terms work-family, family-friendly and work-life balance continue to be used interchangeably in the UK, USA and Ireland.

But despite its popularity, WLB remains far from being a universally understood or unproblematic term. Rather, there is a multiplicity of overlapping concepts that populate the work-life research and policy literatures which have been articulated differently to conceptualise these painful sets of work-life conflicts and WLB as a set of practices and discourses. These include 'negative work-to-home spillovers' and 'negative home-to-work spillovers', understood as the extent to which home demands make one too tired to fulfil the demands of paid work and vice versa (Crompton and Lyonette 2006). This lexicon also includes a variety of references to work-life balance as an objective state of affairs, a subjective experience, perception or feeling, an actuality or aspiration, a discourse or practice, a metaphor for flexible working, a metaphor for the gendered division of labour, or a metaphor for a political agenda (Fleetwood 2007: 352). Table 3.3

Table 3.3   Multiple 'definitions' of work-life balance.

WLB as...	Example descriptions
General discourse	'Work-life balance articulates the desire of all individuals – not just those with family responsibilities – to attain a balance between their paid work and their life outside work, from childcare and homework to leisure and self-development' (Khallash and Kruse 2012: 682).
Aspiration	'WLB ... is about adjusting working patterns regardless of age, race or gender so that everyone, regardless of age, race or gender, can find a rhythm to help them combine work with their other responsibilities or aspirations' (DfEE 2000: 4).
Political agenda	'The work-life balance agenda is where philosophical questions about what is the good life and what is the common good intersect with the political. We need to challenge the centrality of work in our lives, and reconsider the price we pay for our wages ... the economy should be the servant of our needs, not our master' (Bunting 2005: xxvi–xxvii).
Objective state of affairs	'Satisfaction and good functioning at work and at home with a minimum of role conflict' (Clark 2000: 751).
Subjective experience	'Work-life balance is about people having a measure of control over when, where and how they work. It is achieved when an individual's right to a fulfilled life inside and outside paid work is accepted and respected as the norm, to the mutual benefit of the individual, business and society' (The Work Foundation/Employers for Work-Life Balance (Jones 2003).
Practice	'Work-life balance practices in the workplace are therefore those that, intentionally or otherwise, increase the flexibility and autonomy of the worker in negotiating their attention (time) and presence in the workplace' (Gregory and Milner 2009: 1–2).
Alternative working arrangements	'Work-life policies and practices are geared to ... enhance flexibility to increase worker control over the location, place or amount of work, or provide additional instrumental resources such as information and direct services to enable individuals to be able to combine employment with caregiving or other important non-work roles' (Kossek et al. 2010: 2).
Perception/ feeling	'Feeling that work and family/personal demands each make legitimate claims on an individual's time, but the individual cannot control the balance between them' (Tausig and Fenwick 2001: 102).

provides a range of examples of different definitions of WLB commonly cited in the WLB academic research and policy literatures. Against this backdrop, Bunting (2005) has memorably described work-life balance as a 'weaselly term' (p. xviii).

Further criticisms have also been levelled at the constituent 'work', 'life' and 'balance' elements used in WLB terminology, which are by now widely rehearsed in the literature. In summary, these relate to the implicit dualistic suggestion that work is somehow not part of life; that people's lives are only divided between paid work and some other undifferentiated activity called 'life'; its overlooking the possibility that some people might actually like their jobs; its potentially undermining unpaid carework by implying it is just another part of the non-work domain; its (false) implication of work and personal life as mutually exclusive; its overlooking the shifting nature of people's work and non-work involvements and meanings given to these over the lifecourse; its implication that balance and equity are synchronic (relating to now), rather than diachronic (relating to a temporal span); that work is somehow bad and life is good; that a balance has been achieved by some and could be achieved by many more; and that balance is synonymous with 50/50 equity (see, e.g., Rapoport et al. 2002; Mitchell et al. 2004; Moen and Sweet 2004; Gambles et al. 2006; Ransome 2007).

In seeking to overcome these critiques, a range of alternative WLB monikers have been developed which emphasise the mutual constitution of these two spheres and the need for longer-term solutions rather than quick fixes. These include work-life reconciliation, work-personal life integration, work-personal life harmonisation and work-life articulation (see, e.g., Lewis et al. 2003; Lewis and Cooper 2005; Crompton 2006; Gregory and Milner 2009). Nevertheless, these alternative monikers are themselves also not immune from criticism.[15] At the same time, WLB retains a common currency (see Figure 3.1) and instant recognition amongst the high-tech workers, HR managers and employers who took part in this study that none of the alternative monikers matched, hence its preferred usage as a convenient shorthand in this book.

Why All the Fuss? Evidencing the Multiple Negative Outcomes of Work-Life Conflict

Whatever the label used, the societal and moral significance of the successful integration of paid work with other meaningful parts of life is profound, with multiple studies documenting causal connections between poor WLB and increased stress, reduced psychological well-being, deteriorating familial and community relationships, and gendered labour market inequality (Burchell et al. 2002; Gornick and Meyers 2003; Wise and Bond 2003; Greenhaus et al. 2003; Marmot et al. 2010; Burnett et al. 2012 – to name just a few). These effects are a particular cause for concern, given that almost 60% of the 35,000 participants in the most recent Third European Quality of Life Survey (2012) identified difficulties in balancing work and family life (Eurofound 2012: 61–62). Three major areas of impact have been identified.

One of the strongest and most consistent findings revealed in the WLB research literature is the significant relationship between increased work-life conflict and stress-related outcomes amongst workers (Allen et al. 2000; Gatrell and Cooper 2008), including a lack of concentration and lower alertness (MacEwen and Barling 1994); increased frequency of heavy drinking and drinking to cope (Frone et al. 1994, 1997); burnout (Anderson et al. 2002); and higher susceptibility to a range of health problems including heart disease, migraines, stomach problems, depression, emotional problems and musculoskeletal disorders (Dembe et al. 2005). When faced with stress due to time constraints, employees may also forego healthy behaviours including regular exercise, sleep and healthy eating, further compounding the negative health outcomes of work-life conflict. Here, studies have documented connections between overwork and a decrease in home food preparation and family meals (Hochschild 2012) and a reduction in exercise and increase in the consumption of fast foods and convenience meals (Allen and Armstrong 2006), which are widely identified as causing weight gain and increased obesity (Hearst et al. 2012).[16] And beyond the costs to workers and their families, the costs of stress-related illnesses in the workplace are also substantial. For example, work-related stress is conservatively estimated to cost EU businesses €20 billion a year in lost productivity (EU-OSHA 2014). In the UK, work-related stress is estimated to affect 1 in 5 workers, resulting in 13.4 million work days lost per year (Bunting 2005: 17).[17] And in Ireland, commentators have pointed to a potential Irish 'quarter life crisis' phenomenon, in which patterns of work-related stress previously documented amongst workers in their 40s and 50s are now evident amongst some workers in their 20s and 30s (O'Boyle 2006; see also Cullen 2004). Studies in organisational psychology have also identified the impacts of one spouse's work-related stress on the other spouse as 'crossover stress' (Bolger et al. 1989), through which work-life conflict can effectively spill over between companies as spouses increase their own unpaid work in response (Pittman et al. 1996). The implications for organisations are therefore clear: 'work-life conflict can have negative repercussions for employee performance' (Beauregard and Henry 2009: 11).

Second, research has also highlighted the importance of WLB policy intervention as a potential means for improving gender equity in market employment (Wise and Bond 2003; World Economic Forum 2005; cf. Connell 2005). In contrast to enlightened policy rhetoric on WLB as a 'universal issue', studies have also documented the stubborn persistence of gender variations in work-life conflict as women continue to make the greatest compromises to fit paid work around family (Sirianni and Negrey 2000; Moen 2003; McDowell et al. 2005). In the absence of meaningful support arrangements, these challenges are particularly acute for working mothers with young children, when childcare becomes necessary, and as they face a potential loss of seniority, occupational status, income and pension if they take extended breaks from employment (Dex and

Joshi 1999). Consequently, studies have repeatedly documented how gendered work-life conflict constrains female career advancement (Folbre 1994; Dex et al. 2008; Scott et al. 2010), with data suggesting (conservatively) that at least two thirds of the wage gap between men and women reflects women's majority load of family work (Williams 2000). Studies have also documented the negative health-related outcomes for women who 'try to keep up with their male counterparts in the workplace while shouldering the burden of unpaid caregiving' (MacDonald et al. 2005: 65), including stress, burnout and poor mental health. Indeed, it is difficult to overestimate the scale of these problems, given that approximately 90% of women will become mothers at some point during their work-lives (Williams 2000: 9).

Third, the profound societal implications of work-life conflict also include mounting concerns around a 'care deficit', based on a shortage of time and energy to invest in nurturing resilient, secure individuals, families, friendships and communities (Bunting 2005; Gibson-Graham et al. 2013). As work increasingly occupies time previously set aside for families or other interests, studies report increased loneliness, eroding friendship support networks, marital strain, increased likelihood of divorce, and falling quality of life and well-being (Lewis et al. 2003; Johnson 2004; Williams et al. 2008). And all this alongside negative impacts on parent–child relationships (Crouter et al. 2001), shorter periods of breastfeeding for mothers in full-time employment (Lindberg 1996) and delayed parenthood until later in life (if at all) (Frisén et al. 2014). Thus while neoliberal welfare state cutbacks have assumed an infinite capacity on the part of households to absorb the costs of reproduction and care, the reality at the level of workers and families is one of increased feelings of time pressure and reduced health and well-being. Parents also experience anxiety in the face of 'intensive parenting' ideologies which urge greater amounts of time and energy to be directed towards care of children and their leisure activities (Hilbrecht et al. 2008). The overall situation is that:

> The personal and emotional content of home life is becoming more and more concentrated in a relatively small number of activities, such as sharing meals or telling bedtime stories, for which substitutes cannot be purchased. Past a certain point – which our society has yet to define or negotiate – family time cannot be reduced without adverse consequences for all family members. (Folbre and Nelson 2000: 129)

Accordingly, the labour union movement has also emphasised the social importance of work-life balance provision as a means of improving workers' quality of life and combating the increasing work pressures that are destabilising families and households (e.g. TUC 2005; ICTU 2005). The OECD series of reports *Babies and Bosses* also recognises that employer-provided family-friendly working arrangements can increase the living standards of children (OECD 2003, 2005, 2007a).

Enabling Work-Life Balance in Practice? Uneven Geographies of Policy (Non-)intervention

While WLB has risen sharply up the international research and policy agenda, the pressures facing working parents continue to increase, and 'there is no sense that the work-life balance problem has been "solved"' (Gatrell and Cooper 2008: 72). Part of the problem is a lack of clarity around who is responsible for obtaining, or failing to obtain, work-life balance (Fleetwood 2007): as a matter of individual choice at the level of workers (would individuals have obtained a better WLB if they had made better decisions?), firms, or else at the level of government? More generally, while many studies have documented extensively the origins of the WLB problem and argued very forcefully on *why* things need to change, far less is written on *how* to effect positive change, to promote more socially sustainable forms of working *in practice*.

Reinforcing these challenges, the terms in which the WLB debate is cast and acted upon vary nationally (Esping-Andersen 1990, 1999), reflecting different national traditions, cultures, institutional contexts and normative assumptions about gender roles. At one end of the scale, Scandinavian national policy frameworks within the social democratic welfare tradition are best known for fostering more systemic and progressive work-life possibilities through legislation to make it easier for women and men to combine paid work with the rest of life; government concerns to reduce working time for all; and an explicit aim to degender parenthood and caring responsibilities. Developed in partnership between employers, state and labour unions, Swedish workers receive a long period of generously paid parental leave, 2 months of which must be taken by each parent or be lost; extensive publicly provided and state-subsidised childcare; and cash benefits for children. In short, in Sweden the 'care conundrum' is largely resolved by the state (Perrons et al. 2006: 265). At the other end of the policy spectrum are English-speaking countries with liberal welfare regimes (especially the UK and Ireland), where the problem of combining paid and unpaid work has instead long been seen as a private affair (Lewis 2009). Accordingly, mothers are often expected to purchase private childcare arrangements through the market, with employers encouraged *voluntarily* to provide working arrangements that better enable workers to reconcile work, home and family.[18]

Rather than focusing on the intrinsic value of WLB provision as a means for improving the well-being of workers and families, policy-makers in liberal welfare states have emphasised the economic value of WLB provision for firms' profitability, productivity and competitive performance. The policy appeal of this so-called WLB business case is that it communicates in a language which managers understand (Dex and Scheibl 1999: 34), and increases the likelihood of obtaining the commitment of company leadership (Robinson and Dechant 1997: 21).[19] This as part of an explicit effort by governments to win over employers, identified by Lewis (2009) as the most powerful interest group likely to oppose a more widespread provision of WLB initiatives. Based on an appeal to neoliberal claims for the apparent superiority of free market principles, the responsibilities of employers and the state to sustain the social reproduction of labour-power are reduced (see Mitchell et al. 2004; Fleetwood 2007). Instead, caring activities are naturalised (returned to the level of the household) and/or reprivatised (provided through the market for profit) (Brodie 1994; Hochschild 2012). In this way, the contemporary rise of WLB as a set of practices and discourses is part of a broader political project 'to re-establish the conditions for capital accumulation' (Harvey 2005: 19), in which employees' use of 'alternative' working practices is only made 'possible in exchange for a shift of both risk and responsibility – directionally away from the employer and towards the flexible worker' (Dixon 2009: 256).

Unpacking the WLB Business Case and its (Dis)contents

More people today want a life beyond work. Employees can work more effectively if they can integrate their work, families and personal lives in more satisfying ways. This becomes a win-win situation for all involved. (Burke 2005: xii).

While multiple studies have convincingly demonstrated the causal connections between poor work-life balance and increased stress, reduced psychological well-being, deteriorating familial relationships and gendered labour market inequality, many scholars have also come to the reluctant conclusion that employers are unlikely to implement meaningful work-life arrangements unless they can identify bottom-line economic advantages that arise from doing so (Dex and Scheibl 2001; Healy 2004; Hyman and Summers 2004; Kossek et al. 2010). The policy-espoused employer benefits of work-life provision are promoted as improved recruitment and retention; improved morale and productivity; and reduced stress and absenteeism (e.g. Bevan et al. 1997; Employers for Work-Life Balance 2005; IBEC 2002; US Council of Economic Advisers 2010). Reduced operating costs are also highlighted as a function of employees spending less time at work searching for childcare solutions, employees taking fewer sick days per month to care for dependent relatives and the reduced costs of employee replacement (Bevan et al. 1999).

Significantly, these policy claims find empirical support in a range of studies. Table 3.4 lists the main types of employer-provided work-life arrangements. Four common categories are: those that provide greater flexibility in the scheduling of *when* work is done; those that provide greater flexibility in the spatial location of *where* work is done; those that *reduce total* work hours; and policies designed to provide workplace social support for parents, including employer assistance with childcare (Glass and Estes 1997: 294). Table 3.5 evidences a range of studies which have measured systematic relationships between the implementation of

Table 3.4   A typology of employer-provided work-life balance arrangements.

Policy type	Description	Examples
Flexible work arrangements	Policies designed to give workers greater flexibility in the scheduling and location of work hours while not decreasing average work hours per week	• Flextime (flexible beginning or end work time, sometimes with core hours) • Flexplace/telecommuting (all or part of the week occurs at home) • Job sharing (one job undertaken by 2 or more persons) • Annualised hours
Reduced work hours	Policies designed to reduce workers' hours	• Part-time work • Compressed work weeks (employees compact total working hours into 4 days rather than 5) • Term-time working
Personal leave	Policies and benefits that give leave to provide time for personal commitments and family caregiving	• Extra-statutory maternity leave • Extra-statutory paternity leave • Adoption leave • Unpaid leave during school holidays • Guaranteed Christmas leave • Use of own sick leave to care for sick children • Leave for caring for elder relatives • Emergency leave • Study leave • Sports achievement leave
Practical help with childcare	Policies designed to provide workplace social support for parents	• Employer-subsidised childcare – on-site • Employer-subsidised childcare – off-site • Information service for childcare • Workplace parent support group • Breastfeeding facilities • Policy of actively informing staff of benefits available

specific WLB arrangements and different, tangible firm performance outcomes (specifically, increased productivity, improved employee retention, improved recruitment, decreased absenteeism and decreased employee turnover), often using logistic regression modelling. Scholars have also stressed the need to investigate the impact of *bundles* of employer-provided work-life policies and practices to capture a broader set of organisational impacts than can be captured by focusing on individual WLB arrangements in isolation. Most notably, work by Perry-Smith and Blum (2000) shows that firms with a greater range of work-life

Table 3.5 Evidencing the business benefits of employer-provided work-life arrangements.

Work-life policy implemented	Identified benefit at the level of the firm				
	Increased productivity	Improved employee retention	Improved recruitment	Decreased absenteeism	Decreased employee turnover
Flexible work arrangements/Flexible work schedules Scholars tend (differently) to sum flextime, part-time work, telecommuting/flexplace, job sharing, compressed work weeks	Rodgers (1992) Frolick et al. (1993) Seylor et al. (1993) McCampbell (1996) Perry-Smith and Blum (2000) Shepard et al. (1996) Boyer (1993) Dex et al. (2001) Bond and Galinsky (2006)	Hannah (1994) Rodgers (1992) Bevan et al. (1997) Glass and Riley (1998) Bond and Galinsky (2006) Maxwell et al. (2007)	Honeycutt and Rosen (1997) Casper and Buffardi (2004) Maxwell et al. (2007)	Baltes et al. (1999) Dalton and Mesch (1990) Rodgers (1992) Seylor et al. (1993)	Glass and Riley (1998) Rodgers (1992) Batt and Valcour (2003) Seylor et al. (1993) Scandura and Lankau (1997) Dex et al. (2001) Porter and Ayman (2010)
Employer-supported childcare policies	Seylor et al. (1993) Perry-Smith and Blum (2000) Brandon and Temple (2007)	Grover and Crooker (1995)	Kossek and Nichol (1992) Seylor et al. (1993) Thompson and Aspinwall (2009)	Burud et al. (1984) Auerbach (1990) Goff et al. (1990) Brandon and Temple (2007)	Ransom and Burud (1989) Kossek and Nichol (1992) Seylor et al. (1993)
Parental leave Includes maternity and paternity leave (paid and unpaid)	Dex et al. (2001)	Waldfogel et al. (1999) Baker and Milligan (2008)		*Human Resource Management* (1996): paid time off for family illness	Glass and Riley (1998): maternity leave (paid and unpaid)

arrangements in place demonstrate higher levels of organisational performance, market performance and profit-sales growth. Other work has also pointed to a 'management as symbolic action' effect, in which WLB provisions that symbolise an apparent organisational concern and send signals to current and potential employees about a company's values can also provide *intangible* benefits to organisations, regardless of their actual content (Pfeffer 1994).

Problematically, however, 'the gap between enlightened rhetoric about the need for a readjustment in the work-life "balance" and the reality in most workplaces remains disturbingly wide' (Taylor 2001: 15). A major sticking point for positive change is that, despite the popularity of the WLB business case, there remains a relative dearth of empirical evidence to support these claims in practice, such that many employers remain unconvinced (Glass and Estes 1997; Beauregard and Henry 2009). Not least, 'few scholars have demonstrated the mechanisms through which such [WLB] policies function (or do not) to enhance firm performance' (Eaton 2003: 145–146).

Consequently, work-life provision continues to be seen by many employers as too costly and disruptive, especially for small firms (Hogarth et al. 2009; DWP 2010), or as unfairly privileging a small subset of their workforce, especially working mothers.[20] Accordingly, any workers who *do* avail of WLB arrangements are often perceived negatively as less committed and adversely affecting the firm (Scholarios and Marks 2004). Research has also shown that intended WLB beneficiaries are often unwilling *to use* employer-provided work-life arrangements because of perceived negative career outcomes (e.g. Bailyn 1993; Wacjman 1999). This is reinforced through a set of 'Mommy Track' promotion pathways that offer flexibility at the expense of work success (Williams 2000). Scholars have also identified an invisible 'Daddy Track', where fathers use paid vacation for childcare purposes to avoid colleagues perceiving them as less committed to the organisation (Beauregard and Henry 2009) – negative perceptions which have been documented even in the presence of identical worker performance ratings (Wayne and Cordeiro 2003). Requests for flexible working in pursuit of improved WLB are, therefore, often turned down on the basis that they will harm firms' operational effectiveness, invariably by managers 'reluctant to dismantle a system that has framed their logic all of their lives' (Williams 2000: 113). In short, the uncertainty of the WLB business case results in ongoing hardship for many workers and their families as work-life policies and practices are marginalised within organisations, as a privilege rather than a right (Kossek et al. 2010).

In order that work-family policies be integrated into mainstream organisational policies and practices, it is critical that we expand the evidence base that documents their beneficial effects on firm performance (Kossek et al. 2010: 12). At the same time, it is important to recognise that a range of methodological and conceptual shortcomings also limit many of the WLB business case analyses that *do* exist. Three main problems are identified below, which collectively motivate the alternative conceptual and methodological framing developed in this book.

(i) Prioritisation of employer needs at the expense of workers and their families

The first problem with mainstream WLB business case evaluation research concerns a predominant focus of analysis that reflects the needs of employers – productivity, profitability and competitiveness – to the exclusion of broader social *equity* concerns at the levels of workers and their families (Glass and Finley 2002). Within this framework, employee voice is restricted (Hyman and Summers 2004), with the assumption that accommodating those workers who do not currently fit the dominant employee mould represents a cost, not an investment (Rapoport et al. 2002: 42). Indeed, taken to its logical conclusion, a narrowly economic analysis of business imperatives alone would ultimately reject calls for employer-provided work-life arrangements in favour of the simple exclusion of (predominantly female) workers with dependent care responsibilities (Kingston 1990) who do not fit the (predominantly male) 'ideal worker' model defined in terms of presence and commitment (Acker 1998). Indeed, Carnoy (2002) has argued that, in the new economy, 'the very best workers are those who never sleep, never consume, never have children, and never spend time socialising outside work' (p. 143).

Perversely, the business case for WLB has come to legitimise certain types of employee-*un*friendly flexible working practices, which include involuntary temporary working and involuntary part-time working (with loss of pay), zero-hours contracts, Saturday and Sunday working, enforced and/or unpaid overtime, annualised hours, stand-by and call-out arrangements and seasonal work (see Fleetwood 2007: 389).[21] Or in other words, 'flexible working seems to be more concerned with accommodating life to rather demanding and unquestioned working hours rather than one of reorganising work to allow time for domestic and caring responsibilities' (Perrons 2003: 69). A further irony is the willingness of employers to pay to alleviate their employees' stress but their reluctance to provide facilities that might prevent it in the first place (Hogarth et al. 2001). In short, 'unless we consider both business and social imperatives, optimal outcomes cannot be reached' (Lewis et al. 2003).

Herein lies the central relevance of the work-life 'dual agenda' or 'mutual gains' perspective adopted in this book, in which the gendered experiences, feelings and preferences of workers as citizens and parents are analysed *alongside* the economic requirements of the firm from the outset. Crucially, both are accorded equal importance (following Rapoport et al. 2002; Maxwell and McDougall 2004; Wise 2003), to explore how firm performance is enhanced by providing the kinds of work-life arrangements that workers and their families find most useful, rather than those that are simply easiest (cheapest) to implement. In so doing, this dual agenda challenges the asymmetrical power relationship involved in WLB provision that:

> When employers are asked why they do not provide family-friendly working practices (say flexi-time) the typical reply is that 'such arrangements are not compatible

with the nature of the business' (DTI 2004: 68). Imagine the derision that would meet an employee who refused to accept an employer-friendly working practice (say annualized hours) because 'such arrangements are not compatible with the nature of the family! (Fleetwood 2007: 397)

The aim, then, is to demonstrate empirically the ways in which employer provision of *workers' preferred* work-life arrangements can also enhance (or else not undermine) firm competitiveness. This sequencing is crucial to avoid producing an analysis that might otherwise obscure (or worse perpetuate) the genuine WLB problem as lived and experienced by workers and their families. This, whilst also recognising the pragmatic need to frame calls for socially progressive corporate change in a language that managers (and shareholders) can hear (see Froud et al. 2000).

(ii) Narrow focus on output measures of firm performance

Second, most studies of the organisational benefits of work-life provision tend to restrict their focus to quantitative 'output' measures of firm performance: especially workforce productivity (output per employee or output per hour), and its attendant variables of labour turnover and absenteeism (see, e.g., Huselid 1995; Shepard et al. 1996; Dex et al. 2001; Glynn et al. 2002). While these are centrally important measures of 'revealed competitiveness' (Gardiner et al. 2004), of themselves they say little about the underlying sources and determinants of firms' competitive performance, which in the context of the shift to the new economy is increasingly sustained through continuous technological learning and innovation. As documented in Chapter 2, these underlying determinants include firms' abilities to access external information flows between firms (customers, suppliers, competitors), research organisations (universities, training institutes) and public agencies (technology transfer agencies, development agencies), as well as firms' abilities to access less ubiquitous embodied forms of tacit knowledge through networks of repeated face-to-face interaction and by hiring new talent who embody particular forms of tacit knowledge. The underlying determinants of competitive performance also include firms' abilities to reconfigure, transform and use that new knowledge to commercial ends (or 'absorptive capacities').

Firms' capabilities to learn and generate knowledge internally, to access, incorporate and use externally derived knowledge, and to develop and maintain effective problem-solving procedures are not discrete, abstract, disembodied processes. Rather, they are *peopled* by workers with very real social identities, family commitments and extra-curricular interests, who interact on an everyday basis to combine varied skills, competencies, ideas and prior experience to create new knowledge and apply it incrementally in the pursuit of improved economic performance. As such, they are *unavoidably* shaped by the workplace institutional environment of which work-life arrangements form an increasingly important component. Nevertheless, this is an area in which conventional WLB business case analyses remain silent. This analytical silence is perhaps best illustrated in work by Shirley Dex and colleagues of the effects of employer-provided WLB arrangements on firm performance (Dex and Smith 2002). This work builds logistic regression models of these effects based on an impressive 93 separate variables spanning employers' family-friendly practices, firm structural and performance variables, and HR practice and workforce variables. The large diversity of explanatory variables included sets this work apart from other similar studies (e.g. Bloom et al. 2011; Konrad and Mangel 2000; Heiland and MacPherson 2005). However, here and across the board, no learning or innovation variables are included.

(iii) Abstraction of firms from regional economies

A third problematic characteristic of WLB business case evaluation research – also reinforcing its lack of engagement with firms' learning and innovation behaviours – concerns its predominant intra-firm focus of analysis and false abstraction of case study firms from the regional industrial systems of which they form part. Accordingly, while studies have explored the effects of work-life provision *within* firms, we know very little about the effects of work-life provision on the complex relationships and networks of interaction *between* firms. One crucial mechanism for the mobility of knowledge and learning, and hence for the innovative capacities of firms engaged in related economic activities, is the cross-firm job-to-job mobility of highly qualified workers. As employees move between workplaces, comparisons of evolving ideas are made with 'how things are done in other firms' (Henry and Pinch 2000: 198); and new constellations of talent increase the potential for learning synergies as new colleagues expose each other to alternative viewpoints, epistemic habits and technical critiques (Malecki and Oinas 1999). And much more than a once-and-for-all transfer, workers often maintain advantageous learning connections back to their former employers. However, while some studies point to the role of different levels of WLB provision in shaping workers' quit intentions and/or patterns of firm exit (Grover and Crooker 1995; Batt and Valcour 2003; McNall et al. 2009), the work-life research literature generally falls short of exploring the subsequent employer destinations of these workers, and the quality of knowledge and competencies that spill over between firms as a result.

Also emergent from this analytical abstraction, our understanding of how 'place matters' in shaping the performance outcomes of employer-provided work-life arrangements remains limited. Whilst most WLB business case analyses identify the cities and/or countries in which the data were collected, studies to date have typically said very little about how the various WLB performance outcomes identified in Table 3.5 might also be dependent on the characteristics of

the places in which they occur, nor why those observed effects might vary between similar firms located in different places. Multiple WLB business case
analyses using national surveys typically provide no spatial disaggregation of
those data, as if the performance outcomes analysed emerge on some flat surface,
and are somehow immune to local and regional variations in urban form, housing
markets, commuting patterns, industrial structures of growth and decline, welfare support, labour markets, cultures of work and political governance. This tendency is evident across a range of well-known WLB business performance
analyses in the UK, USA, Australia, Germany and France (e.g. Konrad and
Mangel 2000; Perry-Smith and Blum 2000; Dex et al. 2001; Bloom et al. 2010;
Wood and de Menezes 2010). Indeed, it is particularly ironic that this limited
geographical imagination is also evident in studies of the performance outcomes
of teleworking specifically, given that this WLB arrangement is premised on
offering workers great spatial flexibility in the location of their work (e.g. Martínez
Sánchez et al. 2007; Martin and MacDonnell 2012; Coenen and Kok 2014).
These problems are further reinforced by a general lack of regional comparative
research within the work-life research literature. And all this despite widespread
academic and policy recognition of the importance of regions as key loci in the
organisation and governance of economic growth, wealth creation and policy
intervention in advanced capitalist economies. In short, there remains an urgent
need to regionalise the WLB business performance research agenda.

Advancing the WLB Dual Agenda: Sustaining Regional Learning and Innovation

In response to the continued significance and analytical limitations of WLB
business case analyses as identified above, *Work-Life Advantage* explores the
intra-firm and cross-firm learning and innovation advantages that can result
from work-life provision in knowledge-intensive firms. And this as part of a
mutual gains research agenda that moves beyond either/or thinking to consider
both business *and* social imperatives in pursuit of optimal work-life balance outcomes. In summary, the analysis is advanced in relation to four key questions:
What are the common, everyday experiences and outcomes of gendered work-
life conflict amongst knowledge workers and their families? What kinds of
employer-provided work-life arrangements do different cohorts of knowledge
workers find most useful in overcoming those conflicts? How does the uptake of
these worker-preferred work-life arrangements enhance (vs. constrain) learning
and innovation processes both within and between firms? And how and why do
those WLB learning outcomes vary geographically within and between regional
economies? These synergies are explored through a case study of information
technology workers and firms in two high-tech regional economies: Dublin,
Ireland and Cambridge, UK.

An important element of this analysis also explores the possibilities for degendering historical asymmetries of working time, in which the time men spend in paid employment determines what time they have left for caring, whereas the time women spend doing childcare dictates what time is left for paid work (Figart and Mutari 1998). Accordingly, an analytical distinction is made between WLB arrangements that accept the accommodation of working time to conventional gendered divisions of labour (e.g. part-time working, compressed work weeks), and those that challenge masculinist work time models, such as arrangements which offer reductions in the lengths of 'standard' work days, increased work time flexibility (e.g. flexible start and end times), limits on overtime, evening, night and weekend working, and/or employer assistance with childcare (Beechey and Perkins 1987: 107; Figart and Mutari 1998: 462).

The salience of this mutual gains agenda must also be understood in relation to shifting patterns of work-life balance since the onset of the Great Recession in 2008, as a function of increased workloads, heightened fears of job loss, understaffing and unpaid overtime (Kossek et al. 2010; TUC 2009). In this 'competitive and cost-conscious climate' (Batt and Valcour 2003: 190), workplace arrangements designed to help reconcile workers' competing commitments around work, home and family have been far from immune (Galinksy and Bond 2009). Rather, evidence points to a potential retrenchment of work-life provision as some employers seek to cut operating costs (Working Families 2008a; Eccleston 2011), with this recessionary burden placed particularly on female workers with childcare commitments, in a manner that risks the advances made by women in the workplace during the previous two decades (Fawcett Society 2009). Indeed, these burdens and risks also vary geographically as a function of regional differences in recession and recovery, and locally uneven austerity cuts to infrastructures of social reproduction and care within neoliberal welfare regimes. The result is ongoing flux in the relative urban spacing of jobs, housing, education and welfare services which geographically delimit the everyday work-life options available to workers and households.

The broader suggestion, then, is that in the wake of the downturn, calls for altering WLB practices may seem a 'little indulgent' (EHRC 2009: 6), with employer-provided work-life arrangements identified in some corporate narratives as 'the luxuries of a booming economy that cannot be sustained as we seek to recover from recession … and [an] unnecessary hindrance on business performance' (Leighton and Gregory 2011: 11). In response to these combined challenges, commentators have emphasised that 'it would be short sighted to sacrifice flexible working rights on the altar of short-term economic recovery, particularly because they can be complementary' (Leighton and Gregory 2011: 11; see also Lewis 2010). Accordingly, the dual agenda developed in this book – concerned to identify the complementary role of work-life provision in sustaining regional learning and innovation – becomes even *more* salient.

Notes

1   The 'new economy' short-hand term encapsulates the simultaneous rise of feminised services, global reorganisation of business through ICTs, new forms of flexible work and employment, and the decline of the male breadwinner/female caregiver model since the late 1970s (see Perrons et al. 2006).

2   The expansive WLB research literature spans industrial relations, sociology, human resource management, organisational psychology, gender studies, feminist economics, geography, anthropology and urban history.

3   Concerns around the sustainability of working hours from the 1960s onwards also contradict earlier predictions of an increase in leisure time through new technologies. John Maynard Keynes famously predicted a 15-hour working week by 2030 (in his 1930 essay, 'The Economic Possibilities for Our Grandchildren', reprinted in John Maynard Keynes, *Essays in Persuasion* (New York: W.W. Norton & Co., 1963), pp. 358–373). Alvin Toffler also heralded the twenty-first century as an 'Age of Leisure'.

4   Work-life concerns can be historicised even further. Williams (2000: 31) locates the historical separation of work life from home life in the development of the nineteenth-century industrial economy, which separated work and home geographically (into factories) and temporally (into a preset workday), in sharp contrast to earlier patterns which interspersed work and family life. See also Stedman Jones (2014) and Walkowitz (1978).

5   The Dilbert comic strip drawn by Scott Adams appears in over 2,500 newspapers worldwide with more than 150 million readers across 65 countries. As a previous cubicle office worker, Adams receives 350 to 800 emails a day from white collar workers which he uses as content for his strip, to provide a window onto the reality of working life in the new economy.

6   Numerical flexibility is also sometimes referred to as external flexibility. Kevin Doogan (2009) warns against the common conflation of non-standard work with job insecurity (as a 'perverse generalisation' that defines these kinds of work in terms of what they are not (p. 148)). He suggests that despite the rise of part-time contracts in the OECD, recruitment practices are far from casual, and average job tenures often greater than for workers in full-time positions. In short, 'it is essential to separate the temporary from the part-time' (p. 160).

7   Taylor (2013) offers a useful definition of work intensity/work intensification in terms of 'the amount of work/effort required of, and performed by, workers in each unit of time (intensity) and an increase in the amount of work/effort between comparable units of time over time (intensification)' (p. 33).

8   In addition to the Joseph Rowntree Job Insecurity and Work Intensification Survey, Burchell et al. (2002: 5) use multiple national datasets to measure work intensification in the UK, including British Household Panel Survey, Workplace Employee Relations Survey, British Social Attitudes Survey and various surveys conducted by the Chartered Institute for Personnel Development. Relevant data are also contained in the European Survey of Working Conditions.

9   Bupa 'Take a Break' Survey, survey of 2000 full-time UK workers, 6 January 2015.

10   Ireland's National Workplace Employee Survey (2003: N=5,198; 2009: N=5,110 employees) (see O'Connell et al. 2010).

11   Survey of 2,672 Irish workers across multiple industries carried out by employment law consultancy Peninsula Ireland (2008) featured in *The Irish Times*.

12   This rise in dual-earner households in Ireland is underpinned by an increase in the proportion of women in paid employment from 38% to 56% (1993 to 2004). The proportion of men in paid employment also increased, albeit at a more modest rate, from 64% to 76% over the same period (McGinnity and Russell 2007: 326).

13   In the UK and Ireland, these challenges to the male breadwinner model are also manifest through a narrowing of gender inequalities in labourforce participation rates, also visible at the EU level (see Table 3.1). These data are consistent with previous observations by Perrons (2003), who notes that, in the UK, more women (72%), more mothers with dependent children (69%) and more mothers with very young children (58% in 2000 compared with 48% in 1990) are economically active than ever before.

14   These 'messy and fleshy' activities of social reproduction include shopping, cooking, cleaning, daily paperwork, social networking, minding, participating in religious or civic organisations, caring for children and the elderly, and hence mediating with educational, medical and religious organisations (Katz 2001: 711).

15   For example, work-life integration is seen as contentious on the basis that it implies the two spheres must be merged, leading to fears of a contamination or the domination of personal life by the demands of paid employment (Lewis and Cooper 2005).

16   In an analysis of the health and well-being of workers in Silicon Valley, English-Lueck (2010) documents the weight gains of IT workers since moving to the Valley through a sedentary lifestyle and gourmet cafeteria food – known at Google HQ as the 'Google twenty-five' [pounds] (p. 109).

17   At the other extreme of overwork in the UK is the widely reported story of the death of Moritz Erhardt, a 21-year-old German summer intern at Bank of America Merrill Lynch in London, found dead on 15 August 2013 as a result of repeated long hours working and sleep deprivation, including working through the night three times consecutively as part of his internship (e.g. *The Guardian* 2013; *The New York Times* 2013). Other cases of unexpected deaths and apparent suicides amongst young professionals working long hours in the banking sector include Sarvshreshth Gupta (aged 22 years), who had complained about 100-hour work weeks as a first-year analyst at Goldman Sachs' San Francisco office (*The Independent* 2015), and Thomas Hughes, a 29-year-old investment banker at Moelis & Company in New York (*The New York Times* 2015).

18   Reinforcing these problems, more punitive welfare-to-work programmes (workfare) pressure single parents into paid employment, and are not underwritten by the same public level of funding for family leave, childcare and other social infrastructures found in the Scandinavian context (Perrons et al. 2006).

19   Indeed, some labour unions also recognise the appeal of the WLB business case, including Ireland's Services, Industrial, Professional and Technical Union: 'we find it a lot easier to make progress at times when the "business case" and "equality case" coincide ... to harness market forces, as best we can, for the long-term benefit of workers and their families' (SIPTU/Callender 2003: 10).

20   Dex and Scheibl (1999: 24) have broken down these costs in relation to the yearly cost of the policy × number of workers benefiting per year, disruption costs of filling absent colleagues' positions temporarily, temporary reduction in productivity from disruption, and reduced morale amongst employees not benefiting.

21   The 'family friendliness' of different types of employer-provided WLB arrangements depends largely on the amount of control (time sovereignty) that is permitted to the worker, the lived experience of the resultant working pattern and the extent to which the terms of any particular WLB practice are 'dictated' by the employer (Lewis 2009: 106).

Chapter Four
Researching Labour Geographies of Work-Life and Learning in Ireland and the UK

Introduction

Work-Life Advantage explores everyday geographies of work-life, learning and innovation amongst information technology (IT) workers and firms in Ireland and the UK. Debates around work-life balance have come to assume a strong national significance in both countries – this as a function of increased female labourforce participation, new worlds of 'flexible' work, increased household complexity, neoliberal welfare reform and the highest average work hours in the European Union (EU).[1] At the same time, Dublin and Cambridge are both widely recognised as important European 'blueprint' high-tech clusters of interest to policy-makers elsewhere. While both have figured prominently in multiple regional learning and innovation studies over the last two decades, those studies have largely failed to analyse the lived experiences of workers themselves (as opposed to 'labour' as a factor input to production), and have typically divorced these 'regional worlds of production' from regional worlds of social *re*production.

In response, this chapter sets out the national legislative frameworks developed in Ireland and the UK over the last two decades to help workers reconcile conflicts between competing demands of paid work, home and family. It also reflects on the successes and ongoing limits to these interventions as a set of practices and (contested) discourses in each country, as manifest in a series of major surveys of work-life conflict and worker well-being undertaken prior to and after the onset

Work-Life Advantage: Sustaining Regional Learning and Innovation, First Edition. Al James.

of the Great Recession (2008). Against this backdrop, the second part of this chapter compares some core features of Dublin and Cambridge's IT clusters, including IT workforce demographics. Usefully, these data point to the wider worlds of social reproduction from which these knowledge workers are inseparable, in turn conditioned by the national work-life legislative frameworks outlined above. These workers are employed as individuals, but come to work each day as members of families and households. The third part of this chapter details the mixed-methods strategy through which the core empirical evidence base for *Work-Life Advantage* was constructed.

My aim, then, is to understand the intimate everyday connections between knowledge production and social reproduction through the lens of the lives and careers of the women and men who live and work in the high-tech regional economies in Dublin and Cambridge. Also, in recognition of the ongoing need for fuller and more frank methodological discussions within economic geography, this chapter offers guidance for early career researchers looking to pursue similar research. The reality of empirical research is never as neat, straightforward and linear as typically presented (albeit retrospectively) in journal papers.

Work-Life Legislative Provision (and Ongoing Challenges) in Ireland and the UK

Over the last decade, work-life balance has come to assume a strong significance in Ireland as part of an emerging critique around quality of life issues following a period of remarkable national economic growth through the 1990s known as the Celtic Tiger (see, e.g., Kirby 2002; Boucher and Collins 2005; Fahey et al. 2008).[2] During the 1990s, the Irish economy (previously identified as a peripheral member country within Europe) outperformed all other European economies, with GDP per capita growth rates three times the EU average (Roche et al. 2008). One of its most distinctive elements – indeed prompting its renaming as the Celtic Tigr*ess* (O'Connell 1999) – was a sharp increase in female labourforce participation rates, from 34.0% in 1993 to 48.8% in 2003 (CSO 2003; cf. O'Connell and Russell 2005: 20–21).[3] And in contrast to the feminisation of the labour market in other EU countries, this growth in women's jobs in Ireland has been in full-time rather than part-time jobs (see Collins and Boucher 2005: 6). During the same period, Ireland also witnessed increased proportions of dual-earner households, and households headed by a working single parent (Russell et al. 2004). Problematically, however, government encouragement of women entering the labour market was not matched by adequate state provision of childcare to facilitate that transition; childcare has instead been left to families and employers willing to provide it (Collins and Boucher 2005: 7), alongside the expansion of a new marketised childcare infrastructure (Gallagher 2012). Critical commentators in Ireland have therefore questioned the extent to which, in the wake of the

Celtic Tiger, 'we have better lives now that we are economically "better off"' (Fahey et al. 2008: xi; O'Callaghan et al. 2015).

The national legislative frameworks designed to facilitate the reconciliation of paid work and family life in Ireland are summarised in Table 4.1, these largely developed through the Irish Social Partnership process in conversation with employers, trade unions and women's groups, and based on government concerns to promote increased profit and expansion of the economy alongside worker well-being. Strikingly in the Irish context, there are no statutory entitlements to paternity leave; likewise no statutory rights to childcare; and no legal right to work part-time. Accordingly, previous Irish studies have documented high levels of work-life conflict and stress amongst working parents, and a lack of affordable, accessible childcare services for pre-school children and school-going children out of school hours (Redmond et al. 2006: 17). Reinforcing these challenges, at the time research for this book commenced Ireland was identified as having the longest average work hours of *all* EU member states (Cowling/Work Foundation 2005).[4] A major national work-life study (Drew et al. 2002) commissioned as part of the Irish Social Partnership process also documented 86% of senior managers and white collar professionals working longer than their standard contracted hours on a regular basis.[5]

Fast forward a few years and negative changes are also evident in the wake of recession. Here, comparison of the Irish National Workplace Surveys (2003 and 2009) evidences a significant increase in work pressure with the onset of the economic downturn, with over half of workers in 2009 reporting an increase in work pressure over the previous two years, compared with just one third in 2003 (O'Connell et al. 2003, 2010).[6]

In the UK, work-family balance has also become an explicit policy goal of successive national governments over the last two decades, beginning with the election of the New Labour government in 1997. These WLB policy interventions include: increased maternity leave and pay; paid paternity and adoption leave; the right to request flexible working; and new tax credits – all with the aim of facilitating the continuation of paid employment as the desirable norm (Bonney 2005: 400). Likewise, the establishment of the WLB Challenge Fund in 2000 which provided almost £12 million for 400 firms to introduce new working arrangements intended to benefit employees, businesses and customers (Perrons et al. 2006). This is significant at a time when Bevan et al. (1999: x) estimated that UK employees spent an average of 16 hours per week searching for child or eldercare, of which up to 80% was done during work hours. They also calculated that UK employees' use of sick leave to care for dependent relatives cost employers in excess of £50 million per year.

Key features of the UK's national legislative framework designed to promote better WLB are summarised in Table 4.1 and – as in Ireland – include an explicit commitment to demonstrating the 'business case' for WLB. This can be understood as a function of an ongoing UK government unwillingness to interfere in

Table 4.1 Comparing national WLB provision/welfare regimes: Ireland and the UK.

	IRELAND	UK
National welfare regime	• Fits 'liberal' model most closely (Esping-Andersen 1999; Ó Riain 2014) • National Social Partnership mix of market-oriented competitiveness, active labour market policy, incentives to work and sustained welfare provision	• Liberal welfare regime with little labour market regulation (Esping-Andersen 1999)
Work-life balance interventions	• **WLB business case** as part of Social Partnership; National Framework Committee for WLB to realise equality *and* competitiveness benefits • Organisation of Working Time Regulations 2001, Employment Equality Acts 1998, 2004, Forum on the Workplace of the Future 2005	• **WLB business case:** e.g. WLB Challenge Fund (£12 million to 400 firms) • Work and Parents Taskforce 2001, Family Friendly Working Hours Taskforce 2009, Equality Act 2010, Childcare Act 2016, Welfare Reform and Work Act 2016
Flexible working (statutory rights)	• **No legal right to work part-time:** part-time work, job sharing, flexitime and teleworking all at discretion of employer • Part-Time Work Act (2001), Fixed-Term Work Act (2003)	• **Right to request flexible working** through Employment Act 2002; parents with child under 6 and carers of adults (April 2007–); 6+ months' service required • Right to request flexible working **extended to all workers** with 26+ weeks service (June 2014–)
Maternity leave	• **26 weeks** standard + **16 weeks** additional unpaid • Statutory maternity pay: 26 weeks at €230 per week (standard rate January 2014–), no obligation for employers to top up	• **26 weeks** ordinary + **26 weeks** additional • Statutory maternity pay: 39 weeks paid (6 weeks at 90% average weekly pre-tax earnings plus 33 weeks at £139.58 or 90% average weekly earnings, whichever lower)
Paternity leave	• **No statutory provision for paternity leave in Ireland** (employer's discretion)	• **1 or 2 weeks paternity leave at 90% pay**

Parental leave

- **14 weeks per parent (unpaid)** taken consecutively, or in blocks with agreement of employer (children 0–8 yrs)

- Up to **18 weeks unpaid** leave (min 1 wk blocks, max 4 wks p.a., children 0–18 yrs)
- Shared parental leave (April 2015–): set against mother's entitlement to maternity leave (52 weeks). £139.58/wk or 90% average weekly earnings (whichever lower)
- National Childcare Strategy (1998) and SureStart Programme (1998)

Childcare support

- **No statutory entitlements to childcare.** Heavy reliance on playgroups offering short hours of attendance with little public funding

Force majeure (family illness or emergency)

- Employees entitled to leave with pay from employment for urgent family reasons: full pay (less than 3 working days in any 12 consecutive months)

- Employment Relations Act (1999): parents' right to take '*reasonable amounts of time off work*' to deal with sudden, unexpected family emergencies

Statutory annual leave

- **4 weeks paid holiday p.a.**
- 9 public holidays p.a.

- **5.6 weeks paid holiday p.a.**
- 8 public ('bank') holidays p.a.

Sources: Dex (2003); Acas (2009); Lewis (2009); Redmond et al. (2006: 25); www.citizensinformation.ie and www.gov.uk (accessed 26 May 2016).

firms' right to manage. Recent changes to this legislation include the introduction of the Work and Families Act 2006, extension of the right to request flexible working to parents of children under the age of 17 in 2007, additional Paternity Leave Regulations 2010 and subsequent extension of the right to request flexible working to *all* employees after 26 weeks of service in June 2014 (see EHRC 2015). In addition, shared parental leave was introduced in April 2015 allowing fathers to share up to 50 weeks maternity leave with the child's mother. A significant pattern evident in Table 4.1, then, is that the level of national legislative statutory provision to enable WLB in the UK is greater than in Ireland. Nevertheless, ongoing problems of work-life conflict continue to be evident in the UK context, as identified through the UK's First (2000), Second (2003), Third (2006) and Fourth (2011) Work-Life Balance Surveys across multiple sectors, by the Department for Business Innovation and Skills. Notably, in the most recent Fourth WLB Survey of 2,767 workers across the UK, half of all workers reported that they worked overtime in a usual work week due to 'workload demands' (of whom half did this unpaid), which represents an increase over the two previous UK WLB surveys. Additionally, one third of workers surveyed believed that making use of employer-provided WLB arrangements would negatively affect their chances of promotion.

With the onset of the economic downturn, in 2009 the Family Friendly Working Hours Taskforce was established to explore the challenges around improving the availability and quality of family-friendly working practices. Recessionary changes in WLB are evidenced through a comparison of the UK's Workplace Employee Relations Survey in 2004 and 2011, with an increase in the majority managerial expectation that responsibilities for managing work-life conflict rest with workers (from 66% of managers to 77%). These managerial expectations are also evident in the Working Families 'Modern Families Index 2015'. Extending the recessionary evidence base further, the UK's 24-7 Work-Life Balance Survey (Hurst et al. 2009) documented a doubling of absenteeism (from 5 to 9 days per annum) from October 2008 to January 2009, due to increased recessionary workload and stress. And in a 2015 UK survey of over 3,000 mothers by the Equality and Human Rights Commission, over one third did not request their preferred flexible working arrangement because they did not think it would be approved or because they were worried their employer would view their request negatively. In addition, over half of mothers who had their flexible working request approved said they experienced unfavourable treatment as a result (including being given fewer opportunities than other colleagues at the same level; receiving negative comments from their employer or colleagues; feeling that their opinions were less valued; and being given more 'junior' tasks than previously).

Thus, many workers in Ireland and the UK face a complex juggling act between the competing demands of work, home, care and other personal interests and responsibilities. So what, then, are the consequences of these various conflicts for workers' abilities to perform effectively in knowledge-intensive work-places and to engage in everyday learning and innovation processes? To understand these effects, this research focuses on the IT sector, a sector infamous for its long hours work regime and extreme instances of work-life conflict.

Placing High-Tech Regional Worlds of (Re-)Production in Dublin and Cambridge

The IT industry offers a highly pertinent sectoral case for work-life research. Crucially, this is a knowledge-intensive industry at the vanguard of new working practices (Ackroyd et al. 2000; Newell et al. 2002), in which firms compete intensely on the basis of who can bring new products to market quickest, and in which the separation of 'work' and 'life' is substantially more blurred than in more traditional occupations (Hyman et al. 2003; Scholarios and Marks 2004). Studies have also found that IT employees demonstrate particularly high stress levels (Fujigaki and Mori 1997). The central relevance of the IT sector also stems from concerns around a 'leaky pipeline', in which highly qualified female graduates in STEM disciplines are entering this sector in increasing numbers, but subsequently opt out over time as a function of childcare responsibilities and limited workplace support. To be clear, this focus is certainly not to position IT workers as somehow unique. Rather, they offer a particularly visible and accessible vantage point from which to understand contemporary problems of work-life conflict amongst white collar professionals in a wide range of other knowledge-intensive sectors, where embodied skills, ideas and competencies represent the major inputs to, and outputs from, production. Indeed for Richard Sennett, 'this small slice of the economy has a cultural influence far beyond its numbers' (2006: 12).

Previous studies have identified a series of key moments and agents in the development of Dublin and Cambridge's high-tech clusters. The majority of IT firms in Ireland are clustered around Dublin (see Figure 4.1), a growth dynamic that emerged in the 1970s, with subsequent investments by US-owned multinationals from the 1980s onwards (including Digital Equipment Corporation, Microsoft, SAP, Google, Yahoo, Apple and Cisco). This FDI was driven by firms using Ireland as the main base from which to access the European market, particularly for software localisation and technical support (Roche et al. 2008).[7] Other key growth factors include Ireland's EU accession in 1973 providing access to European markets; first-mover establishment of business-oriented undergraduate computer applications programmes at multiple Irish colleges in the 1970s; an English-speaking (and growing) graduate workforce, with many graduate ex-pats returning to Ireland for IT work; favourable corporation tax rates and other grant incentives to IT firms after the Finance Act (1984); and the establishment of the boosterist Irish Software Association in 1978.

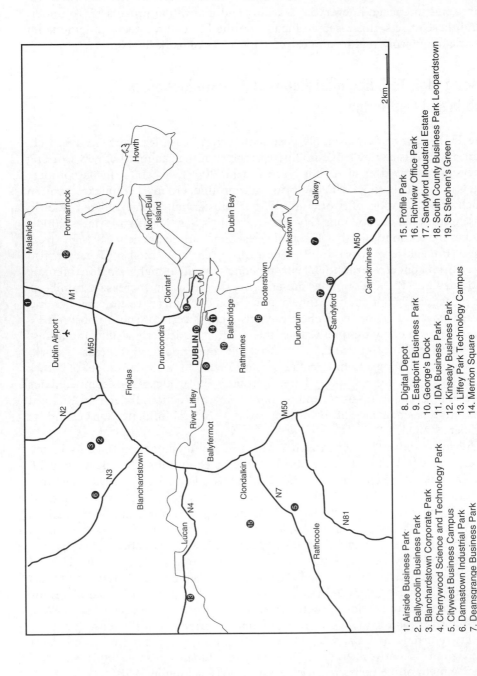

1. Airside Business Park
2. Ballycoolin Business Park
3. Blanchardstown Corporate Park
4. Cherrywood Science and Technology Park
5. Citywest Business Campus
6. Damastown Industrial Park
7. Deansgrange Business Park
8. Digital Depot
9. Eastpoint Business Park
10. George's Dock
11. IDA Business Park
12. Kinsealy Business Park
13. Liffey Park Technology Campus
14. Merrion Square
15. Profile Park
16. Richview Office Park
17. Sandyford Industrial Estate
18. South County Business Park Leopardstown
19. St Stephen's Green

Figure 4.1 Map of Greater Dublin region and key IT sites.

A major study of the Irish IT sector at the peak of the Celtic Tiger growth period identified 30,000 IT workers in 900 companies, with the majority of those workers (83%) and firms (76%) located in the Greater Dublin region (Crone 2002). More recent 2015 data suggest subsequent employment growth with 48,700 IT workers nationally. The IT sector is widely identified as a crucial pillar of the Celtic Tiger, during which time Ireland became the world's largest software exporter.[8] During this growth phase, revenues from the Irish IT sector grew five-fold from €2,189 million in 1991 to €10,150 million in 2000, contributing almost 10% of national GDP despite accounting for only 2% of Ireland's total workforce (Greco 2005: 87). Moreover, during the 1990s, IT employment grew by 15% per annum compared with 6% growth per annum for the Irish national economy as a whole (Arora et al. 2001: 5). The majority of IT firms in Ireland are SMEs, with some larger MNCs also.

The UK's IT industry is spatially concentrated in the East Anglia and south-east regions. As part of this spatial concentration, the growth of Cambridge's high-tech cluster (or 'Silicon Fen') began in the 1970s, following the publication of Cambridge University's Mott Report (1969), which advocated a careful relax-ation of restrictive economic and industrial expansion policies to foster the local development of science-based industries and university spin-offs. Subsequent waves of high-tech growth were therefore enabled through the creation of mul-tiple research parks in and around Cambridge from 1970 onwards to support small start-ups (see Figure 4.2). Other key growth factors include firms' abilities to draw on a supply of highly educated science graduates; Cambridge University's laissez-faire approach to allowing academic researchers to commercialise research and to support spin-off companies; significant R&D activity in embedded labora-tories in Cambridge University through partnerships with companies including Microsoft Research, Intel, Oracle and Toshiba; and the ability of Cambridge University to attract major external grant funding and research centres (see Martin 2006). Consequently, information technology is identified as the region's lead high-tech sub-sector, accounting for the majority (60%) of the 1,525 high-tech firms which collectively employ 53,000 workers (St John's Innovation Centre 2013). As in Dublin, there are relatively few large IT firms in the Cambridge region, with the IT sector instead dominated by SMEs.

Successive cohorts of scholars have drawn on the Dublin and Cambridge cases to advance our theoretical understanding of regional learning and innovation dynamics. Following the first high-profile study on the 'Cambridge phenomenon' (Segal Quince Wicksteed 1985), subsequent studies have identified high rates of collective learning amongst SMEs (Keeble et al. 1999; Huber 2012) and the role of worker mobility between technical consultancies and local high-tech firms in enabling knowledge spillovers (Lawson and Lorenz 1999; Lawton-Smith and Waters 2005). Similarly in Ireland, studies have theorised the Dublin growth dynamic of the Irish IT industry since the mid-1980s in relation to localised knowledge spillovers through worker mobility and cross-firm networks of business

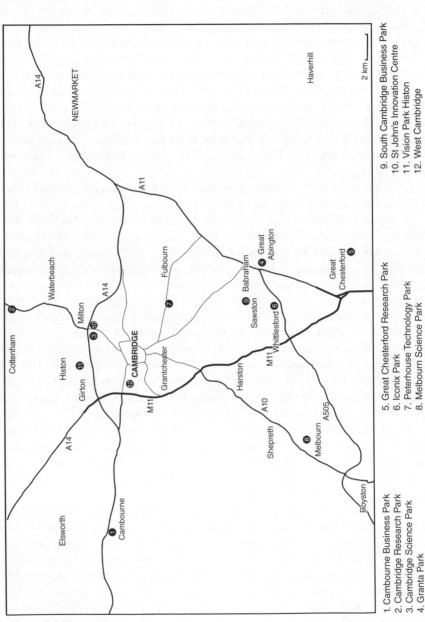

1. Cambourne Business Park
2. Cambridge Research Park
3. Cambridge Science Park
4. Granta Park
5. Great Chesterford Research Park
6. Iconix Park
7. Peterhouse Technology Park
8. Melbourn Science Park
9. South Cambridge Business Park
10. St John's Innovation Centre
11. Vision Park Histon
12. West Cambridge

Figure 4.2 Map of Greater Cambridge region and key IT sites.

contacts (Roche et al. 2008), and their connections to global networks of knowledge production and flows of investment (Roper and Grimes 2005; Breznitz 2012).Yet, largely missing from this body of work in the Irish and UK contexts is any analysis of technologists' work-life experiences beyond mere 'human capital' inputs to knowledge production and exchange processes.[9] This work also lacks any analysis of the active, everyday role of female knowledge workers in (re)making regional geographies of learning and innovation in practice. It is also characterised by the false abstraction of everyday activities of learning, innovation and knowledge exchange from everyday activities of social reproduction. In short, these regional learning studies in Dublin and Cambridge exemplify the kinds of analytical blindspots and biases more widely evident in the regional learning literature identified in Chapter 2.

Identifying IT workers in Ireland and the UK: from faceless 'human capital' to members of families and households

Data on IT workforce demographics are not given in the previous regional learning studies identified above – save for the odd reference to workers' educational qualifications, used to measure the quality of the 'human capital' they offer to firms. In seeking to fill this gap, Ireland's Quarterly National Household Survey (QNHS) and the UK's Labour Force Survey (LFS) offer relevant demographic data.[10] In both cases, IT workers are tracked as part of the NACE Rev 2 Sector J category (Information and Communication), of which three quarters of workers in both countries fall in NACE 62 (Computer Programming, Consultancy and Related Activities) and NACE 61 (Telecommunications).The demographic composition of these IT workers (gender, age, marital status, children, dependants, working hours) is shown in Tables 4.2 and 4.3.

Strikingly, the LFS data show that over half of the UK's IT workforce is married (56%) – this in contrast to popular notions of the carefree Silicon Cowboy. In addition, over one third live in households with young children aged 0–5 years, and just under one third in households with older children aged 6–14 years.[11] Only one fifth of the national IT worker sample is comprised of single men, and the majority (84%) are older than the young twenty-somethings identified by Saxenian (1994).This IT workforce dominance of married workers with children is also evident across earlier phases of the LFS. Similar patterns are also evident in Ireland, as evidenced in multiple waves of the QNHS, with over half of all IT workers married in 2014.[12] One quarter have young children aged 0–5 years, one quarter have older children aged 6–14 years and 44% have one or more children aged 0–17 years. Again divergent from the Silicon Cowboy stereotype, only 27% of the IT workforce is composed of single men, with the majority of IT workers (82%) aged 30 years or older.

Table 4.2   Whatever happened to the UK Silicon Cowboy? IT workforce demographics.

UK NACE Rev. 2 Sector J: Information and Communication (70% of workers in computer programming and telecommunications sub-sectors)	2007	2010	2014
National sample	2299	1310	1457
(east of England subset, region H)	(255)	(117)	(115)
Men	1110	922	1051
Single	400	333	336
Married	**635**	**532**	**633**
Divorced or legally separated	75	57	82
Women	1189	388	406
Single	442	151	160
Married	**590**	**193**	**188**
Divorced or legally separated	157	44	58
In households with 1 or more child aged:			
0–5 years	–	**357**	**499**
6–14 years	–	**429**	**421**
15–17 years	–	110	110
Highest educational qualification			
Bachelors Degree	810	732	467
Masters Degree	–	–	170
Doctoral Degree	10	18	25
Age			
20–29	549	234	233
30–39	**706**	**431**	**426**
40–49	**626**	**348**	**408**
50–59	293	210	264
60+	67	68	110
Size of firm (annual average) (number of persons working in local unit)			
1–10	282	235	290
11–19	143	66	63
20–49	302	173	178
50+	1477	644	**662**

Source: UK Labour Force Survey (Annual Eurostat Datasets), ONS.

In short, these data suggest that IT workers in Ireland and the UK evidence an age profile and parental profile which position them as core intended recipients of the national work-life legislative frameworks in the UK and Ireland outlined earlier in this chapter. Yet these national legislative frameworks around work-life

Table 4.3   Whatever happened to the Irish Silicon Cowboy? IT workforce demographics.

Ireland NACE Rev. 2 Sector J: Information and Communication (79% of workers in computer programming or telecommunications sub-sectors)	2007	2010	2014
	'000	'000	'000
Men	47.5	52.3	57.8
Single	21.5	19.6	22.2
Married	**24.6**	**31.1**	**34.0**
Widowed, divorced, legally separated	1.4	1.6	1.6
Women	21.5	22.6	23.0
Single	12.2	11.6	10.9
Married	**8.4**	**10.1**	**10.9**
Widowed, divorced, legally separated	0.9	0.9	1.2
Have 1 or more child aged:			
0–5 years	12.6	19.6	**20.2**
6–14 years	13.9	18.1	**21.4**
15–17 years	4.9	5.0	5.6
Highest educational qualification			
Third-level non-honours degree	10.4	15.4	12.2
Third-level honours degree or above	**31.8**	**34.1**	**46.8**
Age			
20–29	20.5	16.5	15.3
30–39	25.1	30.8	**31.8**
40–49	13.6	17.0	**20.8**
50–59	7.2	8.9	10.9
60+	1.7	1.4	2.1
Size of firm (annual average) (number of persons working in local unit)			
1–10	–	–	9.6
11–19	–	–	4.0
20–49	–	–	9.8
50+	–	–	**41.0**

Source: Quarterly National Household Survey (QNHS), Central Statistics Office, Ireland. Special Data Extract.

balance (and the employer-provided WLB arrangements which they promote) remain virtually ignored by regional learning and innovation analyses – as if the workers who reproduce regional industrial systems through their everyday work activities function autonomously from them.

Researching Work-Life Geographies of Regional Learning and Innovation in Practice

The core aims of *Work-Life Advantage* are to explore social inequities in the work-lives of similarly qualified workers in doing regional learning and innovation in practice; to make visible the gendered networks of social reproduction and household divisions of labour which reproduce high-quality 'human capital' inputs to firms' knowledge production and exchange processes; to explore how far the everyday workplace practices widely recognised as underpinning learning and innovation might also be socially damaging across the work–home boundary; to identify 'alternative' working arrangements and employer interventions which might simultaneously reduce these negative outcomes for workers and their families and enhance firms' innovative capacities; and to explore the spatial variability of these labour geographies within and between different national welfare regimes. The multi-method research design responds to methodological limits identified in earlier WLB business case analyses, namely: an over-reliance on single-company surveys, a narrow focus on large firm studies at the expense of SMEs, limited measurement of WLB performance impacts through quantitative metrics only, a narrow intra-firm focus at the expense of any cross-firm analysis, and a lack of comparison of firms within and across regional economies.

An online survey of IT employers was undertaken in both regions (summer 2008). Two IT employer databases were built by cross-referencing information from Cambridgeshire Science Parks Directories, Cambridgeshire County Council, Companies House and individual firm websites; and likewise from Enterprise Ireland, the Irish Software Association and Hot Origin. This identified 479 IT firms in Dublin and 504 IT firms in Cambridge by product, market specialisation, address, workforce, key personnel contacts and ownership.[13] The IT employer survey used fixed-choice response questions to document availability and take-up of different WLB arrangements over the previous three years, and managers' perceptions of the impacts of their *total bundles* of WLB provision and take-up over the previous three years on productivity, female recruitment, retention of female employees post-maternity leave, company image to potential employees, workforce diversity and workplace environment for learning and creativity.[14] To judge the credibility of these managerial perceptions, the survey also compared them with changes in common metrics of firm competitiveness (2007 and 2004 datapoints): revenue growth, productivity, % female workforce, workforce attrition. The final employer survey dataset comprised 150 IT firms employing 8,068 workers in the Dublin/Cambridge regions (average 19.3% female workforce) and with combined revenues of £827 million.[15] These are predominantly small firms (50 employees or less), 90% with annual revenues up to £5 million. This SME dominance is consistent with the larger populations of IT firms in both regions.

Alongside the employer survey, 68 in-depth interviews were undertaken through successive periods of fieldwork over three years between 2006 and 2008. Early participants were identified using the IT employer databases described above, with a subsequent focus on firms which had identified positive learning and innovation outcomes as a result of their bundles of WLB provision. In total, interviews were completed with 53 IT professionals: 30 in Dublin across 15 different IT firms and 23 in Cambridge across 14 different IT firms (these firms a deliberate mix of large multinationals with over 250 employees and SMEs). To increase the potential transferability of the research findings, alongside software engineers and programmers, this purposive interview sample also targeted a range of 'semi-professional' (Collins 1990) knowledge workers – including systems management, marketing, sales, human resources and logistics – commonly found in other sectors, and whose daily routines and interactions also form a crucial component of firms' capacities for learning (see Maskell and Malmberg 1999; Benner 2003). The worker sample included 34 parents (including 11 fathers) typically in dual-career couples.[16] But to challenge the divisive assumption that WLB is the exclusive preserve of working parents, the research participant sample was also constructed to include 19 workers with varied family, household and extra-curricular responsibilities and commitments beyond childcare.

Interviews lasted 1–2 hours and explored work-to-home and home-to-work spillovers; the relative utility of different WLB provisions in reducing those spillovers; differences between formal work-life provision and informal support for using those in practice; how the use of different WLB provisions (by workers, colleagues and managers) impacts on *what, how, where, when,* and *with whom* work is done; participants' cross-firm job-to-job mobility paths; and the role of work-life factors in shaping those trajectories.[17] Crucially, these interviews also allowed me to get at what was *not* done (particularly at work-life constraints on worker behaviours), and thereby to produce more nuanced understandings of the gendered work-lives of technology professionals, and of their respective households and employers as constantly moving targets. Participants were also questioned on their before-and-after experiences of significant discontinuities in their use of different WLB arrangements.

An additional 15 interviews were undertaken with labour organisers, economic development agencies, women's IT organising initiatives and local academics to explore further the everyday mechanisms that might underpin any WLB learning outcomes identified in the employer survey, and changes over time. These included the Irish Equality Authority, SIPTU, Irish Congress of Trade Unions, Irish WLB Network, Irish Business Employers' Confederation, National Centre for Partnership and Performance, Economic and Social Research Institute, Girl Geek Dinners, Women In Technology and Trades Union Congress South-east Region. All interviews were recorded and transcribed in full.[18]

Third, an online survey of IT workers was undertaken in both regions (autumn 2008) to document the wider prevalence of work-life phenomena identified at interview. Specifically, the IT worker survey measured workers' demographic backgrounds and household situations; frequencies of experiencing different dimensions of work-life conflict; workers' uptake of specific WLB arrangements in current employers; career history; and how work-life concerns had shaped their decision-making around cross-firm job-to-job mobility, and the shape of those trajectories. The survey participant sample was recruited initially using contacts introduced by participants in the in-depth interviews, and expanded using email listservs administered by three women's IT networking organisations: Girl Geek Dinners (UK), Women In Technology (UK) and Women In Technology and Science (Ireland).[19] In total, 162 questionnaires were completed. It proved difficult to recruit men (n = 9), reflecting powerful, ongoing constructions of WLB as a 'female-only' issue within the IT sector. The largest professional cohorts within the IT worker sample are technical engineers (39%) and technical managers (29%), who along with senior managers and research account for 89% of the worker survey sample. The major household cohorts within the IT worker sample (2008) are working mothers with children in dual-earner households (29%), women without children in dual-earner households (32%), single women without dependants (26%) and single women with children (14%).

The fourth component of the methodology involved participant observation in a series of in-house work-life balance training events held by a large, multi-site IT employer in Dublin; an industry/government/trade union round table debate on WLB hosted by Orange Telecommunications at the (then) DTI; plus work-life seminars, networking events and mentoring sessions held by two prominent female-dedicated labour market intermediaries seeking to empower female technologists in the IT sector: Women In Technology and Girl Geek Dinners.

Fifth, to extend the 2006–2008 analysis into the subsequent recessionary period, an additional online survey of IT workers was undertaken in November–December 2010. Research participants were recruited from the pre-recession phase of fieldwork, and through the Girl Geek Dinners and Women In Technology (UK) and WITS (Ireland) email listservs. This recessionary worker survey yielded a dataset of 147 IT professionals (139 women, 8 men), documenting how workers' everyday experiences of work-life conflict, and availability and use of employer-provided WLB arrangements had changed through the economic downturn. It also documented the extent to which workers' WLB concerns continue (or not) to shape patterns of cross-firm job-to-job mobility. As in the earlier survey, the largest professional cohorts within the IT worker sample are technical engineers (46%) and technical managers (21%), who along with senior managers and research staff account for 86% of the worker survey sample. The major household cohorts within the 2010 IT worker sample are women without children in dual-earner households (35%), women with children in dual-earner households (19%) and single women without dependants (23%).

The analytical strategy connects managers' and workers' perceptions and lived experiences of the learning benefits of WLB provision within their respective firms with measured changes in firm performance across a range of metrics over the same timeframe. The strength of this strategy rests on the principle of convergence: that when multiple data sources (each with their own limitations) are brought together and provide similar and/or complementary findings, the credibility of the analysis is increased. Member checking was also used to gauge the credibility of my evolving ideas and theories with IT technologists and managers in both regions. Much of the information upon which this book is based has been gleaned through personal, albeit formalised, exchanges. Participants often disclosed sensitive information and I have therefore not named names in the analysis itself, instead maintaining rigour by describing the relevant participant's work-life positionality as far as possible within the boundaries of anonymity. I also refer to IT firms by pseudonyms to protect the confidentiality of my sources. That analysis is developed in depth over the next three chapters and incorporates verbatim quotations throughout as part of a deliberately polyphonal and collaborative text, in which research participants articulate their daily struggles to reconcile competing responsibilities and commitments of home, work and family in their own words. In so doing, they vividly demonstrate the inseparability of regional worlds of knowledge production from regional worlds of social reproduction.

Notes

1   A major study by the Work Foundation (Cowling 2005) identified Ireland and the UK as having the longest hours work cultures in the EU (p. 24), based on representative survey data from 21,000 workers across 15 EU core economies.

2   During the period 1994–2002, Irish national GDP grew by an average of 8% per annum, with concurrent growth in the labour market of 4.5% per annum (see Kirby 2002), enabling Ireland to overcome the balance of payments deficit that characterised its weak economic performance in the 1980s. This high growth phase is usually dated to the fiscal austerity programme implemented by the minority Fianna Fáil government which took office in March 1987 (Kirby 2002: 11). Commentators have explained the subsequent Celtic Tiger growth phase in terms of Ireland's success in attracting foreign direct investment (FDI), state investment in education, macroeconomic stabilisation, EU funding, and expansion of tourism and indigenous industry (Breathnach 1998; Grimes 2003; cf. Breznitz 2012).

3   Key factors underpinning these increased female labourforce participation rates in Ireland include changing fertility patterns, higher educational achievement and the removal of legislative barriers on female employment.

4   The UK has the second longest average work hours of all EU member states with figures of 5.8% and 3.9% respectively (Cowling/Work Foundation 2005).

5   Ireland's high-profile *Off the Treadmill* WLB survey (Drew et al. 2002) was based on a national sample of 912 employers and 1,006 workers across multiple sectors.

6   The Irish National Workplace Surveys 2003 (n = 5,509) and 2009 (n = 5,110) carried out by the Economic and Social Research Institute with the National Centre for Partnership and Performance.

7   The rapid ascendancy of the IT industry from the 1970s onwards was driven by the emergence of cheaper personal computers and its challenge to established mainframe computers. For useful commentaries on the historical growth of the Irish IT industry see Crone (2002), Heeks and Nicholson (2004), Sterne (2004) and Roche et al. (2008).

8   Other leading industrial sectors that underpinned Ireland's Celtic Tiger growth phenomenon include computer hardware manufacturing, pharmaceuticals and medical instruments, and all are dominated by FDI by foreign-owned MNCs.

9   But see studies by Ó Riain (2000) and Gray and James (2007) as important exceptions to this general pattern.

10   The UK Labour Force Survey is the largest regular social survey in the UK and is undertaken quarterly (annual figures are derived from Q2 of each year) by the UK Office for National Statistics (ONS). The LFS target population is based on the resident population in the UK, with a sample of 41,000 responding households representing 0.16% of the GB population. Crucially, the 2014 dataset includes a national cohort of 1,457 IT workers (of whom 115 are based in the east of England).

11   Given the household phrasing of the Eurostat data for the UK, whilst we can assume that most of those children represent dependants of these IT workers, in some cases they might also represent their younger siblings.

12   Ireland's QNHS is a national-scale survey of 26,000 households per quarter by the Central Statistics Office.

13   IT companies were defined as companies which develop software products or systems or software development tools for subsequent sale; companies which provide services directly to the design and/or development of software systems; companies involved in the localisation of their own products or third-party products; companies involved in the development of programmes or software systems for incorporation into dedicated hardware devices; and companies providing technical training in systems analysis, design and programming. Hence, the IT employer survey excluded resellers/distributors of third-party software packages, data entry/processing companies and computer maintenance companies.

14   Managerial perceptions of changes in firm performance as a function of firms' total suites of work-life provision and take-up were measured on a 5-point Likert scale: much better, better, no change, worse, much worse.

15   The IT employer survey dataset (2008) comprised 74 firms in Dublin (89% indigenous, 11% foreign-owned) and 76 firms in Cambridge (95% indigenous, 5% foreign-owned). Dublin revenues were converted from € (€1 : £0.78), 1 May 2016.

16   Although it was not a deliberate recruitment strategy, my interview participants did not include any same-sex couples.

17   One issue concerned the irony of asking stressed, busy workers to take time out from their hectic schedules to meet for interview, potentially confounding the very problem under investigation. To minimise these impacts, interviews were often held over lunch (one small way of offering participants something in return for their valuable time) or else at times and places identified as least disruptive by research participants.

18 Analysis of the interview transcripts was carried out through detailed coding and cross-comparison of coded transcripts to draw out key themes, commonalities of experience and sources of difference with the aim of building theory iteratively.

19 Girl Geek Dinners was established in the UK in 2005 targeting women in the IT sector (industry, schools, universities) and now has chapters in over 100 cities and 38 countries worldwide (see http://girlgeekdinners.com/). Women In Technology (established 2004) is a UK networking forum of over 7000 female technologists (see http://womenintechnology.co.uk). Women In Technology and Science (WITS) Ireland is a forum of over 500 women (see http://witsireland.com).

Chapter Five
Juggling Work, Home and Family in the Knowledge Economy

Introduction

> The active units behind the formation of new knowledge are 'epistemic communities', simply defined as groups of knowledge-driven agents linked together by a common goal, a common cognitive framework and a shared understanding of their work. (Cohendet et al. 2014: 930)

> If we were to change the central character in our economic story from the radically autonomous, isolated agent … to the socially and materially situated human being, what effect would this have? (Nelson 1992: 118)

Over the last two decades, geographers have analysed in ever finer detail how the spatial co-location of firms in regional industrial clusters helps foster conditions conducive to learning, innovation and economic competitiveness. This research agenda has yielded important insights into the socio-cultural foundations of regional economic advantage and fundamentally shaped the content and focus of regional development policies worldwide. Nevertheless, as identified in Chapter 2, there remains a series of peculiar biases and analytical blindspots within this expansive literature. Crucially, despite the apparent prominence of labour, workers are typically conceptualised as a passive factor input to production, similar to capital, equipment or raw materials. In contrast, workers' lived experiences of *being used* as human capital inputs to knowledge production and exchange processes tend to drop out of these analyses. Reinforcing this narrow conceptualisation of labour, regional learning analyses have typically failed to locate activities

Work-Life Advantage: Sustaining Regional Learning and Innovation, First Edition. Al James.
© 2018 John Wiley & Sons Ltd. Published 2018 by John Wiley & Sons Ltd.

of knowledge production within the wider networks of social reproduction and care that enable *socially sustainable* innovation and worker learning in practice. That is, as if workers act autonomously rather than as members of families and households, and as if the knowledge production activities in which they engage on a daily basis 'can be separated analytically and ontologically from their wider existence' (Castree 2007: 859). This is problematic given that labour's 'capacity to do work is not innate; it is socially created and sustained' (Block 1990: 75), typically with women bearing the brunt of these tasks. Thus whilst feminist geographical deconstructions of 'life's work' (Mitchell et al. 2004) have documented the vital economic role of hidden household caring activities in sustaining the formal economy – also transforming the very notion of 'the economic' itself (McDowell 2000) – there remains a general lack of engagement with these debates within the regional learning and innovation literature. Overall, then, 'the wider consequences of economic change or firm competitiveness for the well-being of people in places are correspondingly neglected' (Perrons 2001: 208–210).

Moving beyond a focus on firm-centric conceptions of 'labour' or 'human capital' as mere factor inputs to knowledge production, the alternative analysis developed in this chapter explores workers' everyday experiences of *being used* as labour. In contrast to the productionist focus of most regional learning studies, this labour geographies analysis also makes visible the gendered networks of social reproduction which serve to recreate high-quality 'human capital'. It also explores the ways in which the very working arrangements and workplace practices that are widely heralded to underpin the competitiveness of knowledge-intensive firms through time-based competition can be socially damaging and unsustainable, often with negative implications for workers' health, well-being, family relationships and quality of life. These impacts are explored alongside a focus on the managerial relationships and peer expectations which differently enable knowledge workers to reconcile competing commitments around work, home and family, and the types of employer-provided WLB arrangements that different worker cohorts find most useful.

Within this framework, the chapter proceeds as follows. The first section explores the everyday geographies of work-life conflict amongst IT workers and their families (so-called negative work-to-home spillovers), juxtaposed against previous ethnographic and popular media portrayals of high-tech work-lives. In contrast to enlightened policy rhetoric around WLB as a 'universal' issue, the analysis demonstrates how everyday struggles to juggle work, home and family are particularly acute for female IT workers with young children who continue to undertake the majority share of household labour and childcare. Likewise the challenges of some men who are taking on a larger share of the 'second shift' alongside IT work. The analysis also identifies negative outcomes of these conflicts, not simply for workers and their families, but also for firm competitiveness (negative home-to-work spillovers), through a focus on highly qualified female IT returners taking compromise jobs which reduce work strain but ultimately under-utilise their skill-sets, knowledge and experience; indeed some women subse-quently quitting IT employment altogether, resulting in a loss of embodied skills, training investment and accumulated experience to the IT industry.

In response to these difficulties of *doing* regional learning and innovation in practice, the second part of this chapter explores workers' preferred employer-provided WLB arrangements to meaningfully reduce those tensions. These pref-erences are also compared with the kinds of WLB arrangements most commonly provided by IT employers in Dublin and Cambridge, understood in relation to ongoing barriers to WLB provision and uptake. The latter part of the chapter explores varied changes in IT workers' experiences of work-life conflict through the recession and subsequent period of austerity in Ireland and the UK, and of the subtle complexities of WLB rollback by technology employers.

Populist Portrayals of High-Tech Work Life: Genius Warriors, Go-To Guys, and Boys' Own Heroes

> The significance of the ways high-tech managers and workers conduct their lives reaches far beyond the limited, highly specialized arenas in which they work. The world of high-tech has come to be seen … as something of a microcosm in which the complex, changing cultural realities of work in general – and particularly professional and knowledge-based work – are reflected. (Kunda 2006: vii)

> For some time, I looked for an icon of Silicon Valley … the photo editor of the San Jose Mercury News pointed me to a photograph taken by Meri Simon, titled 'Sleepless in Silicon Valley'. The photograph is of a programmer sleeping on the carpet in his cramped cluttered office, head jammed under his desk, gray blanket draped over his fragile torso … the programmer was none other than David Filo … cofounder of Yahoo! (Bronson 1999: xiv–xv)

High-technology industries have been widely identified as at the vanguard of new working practices and – for some commentators at least – the very epitome of what work is: providing templates for how best to manage it, and images of who workers are. The high-tech banner includes multiple high-profile sectors in which knowledge represents the core input to, and output from, production – including biotechnology, semiconductors, aerospace, medical instruments, electronics and telecommunications. Ultimately, however, it is the information technology industry which seems to evoke the most enduring and infamous images of the world of high-tech. Accounts of the fast-paced, high-pressure IT work environ-ment, 'where the only constant is change' (Kunda 2006: viii), celebrate engineers' total devotion to work and a love of technology, emergent from the hobbyist backgrounds of earlier computer visionaries such as Steve Jobs, Steve Wozniak, Bill Gates and David Filo. Romanticised heroic portrayals of Silicon Cowboys engaged in long work hours in pursuit of big rewards depict work itself as intrinsically satisfying, with high-tech work environments identified as a natural extension of college (Sharone 2002), in which the boundaries between work and social life are significantly blurred. Likewise, common portrayals of a hegemonic occupational masculinity within the IT sector, with terms such as 'digital warriors' and 'go-to guys' (Cooper 2000), 'IT road warriors' (Ahuja et al. 2007) and 'Boys' Own heroism' (Henwood 1993) used to encapsulate masculinist ideal worker norms, in which exhaustion, presenteeism, stamina, virility and work addiction commonly convey workers' depth of commitment. The result, no less, is 'a new masculinity for the new economy', in which young wannabes become 'real men' by putting work first, and getting the job done no matter what (Cooper 2000). In short, work-life balance becomes positioned as a non-issue, because work *is* life for these young men.

These popular conceptions of IT work and workers can be traced back to a series of key studies in the 1980s and 1990s which have sought to codify high-tech work cultures in terms of the behavioural displays that are expected and rewarded, workplace rituals, and visible artefacts that embody deeper underlying values and beliefs – or more simply 'the way we do things around here' (Deal and Kennedy 2000: 4). These ethnographic portrayals include *The Soul of a New Machine* (1981) by Tracy Kidder, which explored the work-life experiences of software engineers at IBM in the early 1980s. Here work is described as being 'like a drug' (p. 90) amongst young male programmers, who 'regularly attended these sessions of midnight programming, ... began to ignore their girlfriends and eventually lost them for the sake of playing with the machine all night' (p. 96).[1] Similar to this is Fred Moody's (1995) heroic portrayal of the work-lives of computer programmers at Microsoft Corporation which underpin the company's position as 'one of the most dramatic success stories in American history' (p. xvii). This success is built on the backs of young programmers:

> Javin Gammill, a software development engineer at the Microsoft Corporation, is in the final stages of a years-long project. He has been working sixteen hours a day, seven days a week for so long that his supervisor has been pleading with him for weeks to take a day off. (Moody 1995: 1–2)

These portrayals also include Pascal Zachary's (1994) *Showstopper!* account of 'code warriors' at Microsoft working to roll out the next version of Windows NT, which also points to the total commitment and personal life impacts of these kinds of classic working patterns in the IT sector:

> Cutler took Digital by storm ... He was completely absorbed in his work. By then, his first marriage had broken up and his second one was headed toward the shoals. He was, by his own admission, no family man. ... Cutler was the classic programmer. (Zachary 1994: 18)

In the final push for the July release … He worked every day during the month of June, some days for as long as twenty hours. He took most of his meals at Microsoft … Fogelin never went home for thirty days during the height of the push. (Zachary 1994: 216–217)

Similar images of work devotion are evident in Robert Cringeley's (1996) *Accidental Empires*, distinctively subtitled 'How the boys of Silicon Valley make their millions, battle foreign competition, and still can't get a date'. Likewise the images of high-tech work life that endure through more recent media statements, for example how 'forward-thinking' high-tech companies, including Google, Facebook, Yahoo and Dropbox, ensure that technologists are not distracted by everyday 'life' needs by variously providing breakfast, lunch and dinner, shower facilities, dry cleaners, gyms, yoga studios, bike repair, candy shops, dentists, doctors, video arcades, barber's shops, music studios, bowling alleys and mail facilities on-site (e.g. Stangel 2013); or headlines which proclaim that in 'Silicon Valley Work-Life Balance is for Losers, Not Closers' (Forbes 2012). The long-standing popular discourse surrounding the demographic and working lives of IT workers is therefore one of young, carefree Silicon Cowboys without family commitments and responsibilities who instead form their primary identities around work, and who regularly work long hours to complete large workloads in short periods of calendar time.

Significantly, these populist portrayals of high-tech masculinist work culture also find support in a range of academic studies. In an early intervention, Rosabeth Moss Kanter (1983: 203) suggested that high-tech work offers a high, a sense of community and a total commitment for many workers that is dramatic, exciting, engrossing and attractive.[2] This total commitment is also evident in Seán Ó Riain's more recent (2000) ethnography of software developers in Ireland in which his tight-knit, all-male development team takes pride in their hectic pace of work, with regular periods of pre-deadline team introversion, and for whom '"do what needs to be done to get this specification working by the deadline" is the broad task of the team' (p. 16). Indeed, some three decades on from Kanter's study, the 'live to work' ethos of high tech is also evident within my own interviews, including some technologists whose work-life balance concerns were to accommodate *more* work rather than less, and whose managers 'you know, almost have to kick them out of the office sometimes'. Participants variously described high-tech work as 'seductive', such that 'there's not enough hours in the day for me to work. Like I would happily slow the rotation of the earth so that I could have an extra couple of hours, if that was possible' (founder and CEO, father of two pre-school-age children, IT SME, UK SE region). The family impacts of these patterns of working were also recognised:

'I think like all macho men, you subjugate the needs of your family. One bank holiday weekend, I was in work for something like 18 hours on Saturday, 14 hours

on Sunday and 12 hours on Monday, and my wife nearly drove off and left me …
But everybody knew what we were trying to achieve and we had brilliant fun. We
actually put beds there and some people didn't go home, they just slept up there for
maybe six hours, we brought pizzas in, it was great fun.' Director of Human
Resources, male, two children (middle school), IT MNC, Dublin

Participants also described the role of employer-provided meal facilities at some
companies in enabling repeated long hours of working, in corporate environments
that are deliberately designed for that purpose:

'I work long hours. So, I might take an hour out for lunch, I might take an hour to
go out swimming and then I come back, have ice cream and then decide I need to
do some work, and then it's 9 p.m. by the time I go home. But that's something that
is encouraged. The fact that we have breakfast, lunch and dinner here makes it easier
to live at work'. Software Engineer, female, IT MNC, UK SE region

The general consensus, then, posits that high-tech jobs offer an exaggerated
form of a broader historical trend in which jobs are structured around the concept
of the ideal worker, usually defined in terms of presence-as-commitment
(Gambles et al. 2006), or over-work as 'a badge of honour' (Gershuny 2005) and
without obligations that would limit the time available to devote to paid work.
On this basis, 'the ever-present and unrelenting pressure to produce and produce
quickly' (Matthews 2003: 207) means that the vast majority of high-tech jobs are
not designed to be combined with other duties such as caring for others, managing
a household, breastfeeding and childrearing. Problems then arise for the many
workers that *do* have outside obligations, such that many women become deval-
ued as they don't fulfil these work ideals (Peterson 2007).

In some ways, this body of research offers a useful counterpoint to commonly
abstract analyses of epistemic communities in which the identities and motivations of 'knowledge-driven agents' (Cohendet et al. 2014) as real people drop out
of the analysis. At the same time, however, the 'industry standard' identification
of knowledge workers and their work-life priorities within these accounts is
woefully narrow: typically, young men whose lives and loves revolve only around
technology work and whose apparent solution to problems of work-life conflict
simply involves a rejection of family in favour of paid work. Against this backdrop,
the analysis developed in the rest of this chapter suggests a more nuanced high-
tech work-life reality, which also recognises male engineers with significant
childcare responsibilities who are no longer able or willing to buy in to these
heroic types of high-tech work culture – if, indeed, they ever did in the first place.
Likewise the multi-dimensional work-life experiences of female IT professionals,
who in Ireland and the UK account for 21% (or over 276,000) of IT specialists,[3]
but whose voices remain largely silent within the expansive regional learning and
innovation literature.

Back to the Future: When the Silicon Cowboys Hang up their Spurs...

In contrast to dominant accounts of Silicon Cowboys engaged in long hours of working driven by 'a love of technology' and 'the sheer thrill of work' at the expense of personal life, my interviews identified a more complex and contradictory set of worker experiences and sources of time pressure and work-life conflict amongst male technologists along three dimensions. In combination, these serve to position popular conceptions of the Silicon Cowboy as historically contingent, and/or over-played. First, whilst not denying the reality of regular long hours for some research participants (figures of over 60 hours per week were not uncommon), the majority typically work 40–45 hours per week (see also Davies and Mathieu 2005; O'Carroll 2005).[4] The results also suggest that everyday experiences of time pressure in the UK and Irish IT sector are not always a function of long hours *per se* but of the temporal variability of work hours over the course of software product development, through which firms seek to bring new products to market quickest. This variability emerges at the project level, in which workers 'could have a 6-month project, could be 11, 18, 23 months, there are no rules ... the closer you get to that date, they work longer, they work weekends, it's part of the job' (Director of Software Development, female, IT MNC, Dublin). And just to clarify those long hours in no uncertain terms, 'you look at releases going out into the marketplace and we would have people who were working 24 hours at times, wrongly' (Director of Human Resources, male, 2 children, IT MNC, Dublin).

In short, the volume of IT work is uncertain, with unanticipated complications along the way also demanding coherent solutions rather than 'shortcut' patch fixes. The result is significant temporal variability of workflow relative to product deadlines, which are fixed and certain. Other commonly cited sources of time pressure in the IT sector include the need for rapid response to client system emergencies, communication between international work team members in different time zones requiring extensions to the working day, being on-call at evenings and weekends for troubleshooting, the necessity of weekend travel, and the continual need to maintain skill-sets 'after hours' in the context of a continually changing industry. The interviews also made clear that problems of time pressure amongst IT professionals are common across larger MNCs as well as smaller start-ups. This temporal variability of IT work generates significant work-life conflict as many of these men try to juggle those demands with personal life responsibilities and commitments whose timings are instead fixed and regular, including: daily school runs, daily picking up partner from work, weekly Cub Scout leadership, weekly amateur team sports fixtures, parents' evenings, assemblies, homework and evening classes. Tensions also arise in relation to the fixed schedules of school holidays and term times. Additional difficulties are also recognised in relation to post-separation parenting, although none of my interview participants had direct personal experience of this.

Second, previously young, single male developers commonly identified significant shifts in work hours and their identification with a 'live to work' attitude over the lifecourse, coincident with marriage, children and changing life priorities. Or as one male CEO of an IT start-up put it, 'AD, after delivery, and BC, before child'. In short, several of these men had consciously rejected former Silicon Cowboy versions of themselves, typically as a function of fatherhood:

'Software development does get a reputation as being a young man's game. There is a continuous onslaught to absorb new stuff and new languages. But as you get older and have kids you (a) can't, and (b) may not want to spend all your time in front of a computer geeking out and doing highly technical work. The guys who've had kids along the way have definitely altered, they've become more judicious about how they apply their time to make the biggest impact'. CTO and founder, male, IT SME, Dublin

'I don't do anywhere near the amount of hours I used to do. I mean in the space of three years that I worked at [previous IT company] I worked five years (I kept my own personal timesheets). I was pulling *unbelievable* shifts, sometimes doing all-nighters, sometimes working until two or three in the morning, like my wife would come in and fall asleep in the seat next to me. Which isn't so bad when you're young, single, free or otherwise, but getting married, people tend to do much less. They've got a bit older, got a bit wiser, they've settled down, they've got a family. People just say, "We're not going to do it anymore".' Software Developer, father of two pre-school-age children, IT MNC, Dublin

In this way, and paralleling Massey's earlier (1995) study of male IT engineers, some men are seeking to challenge dominant masculinist notions of the 'ideal IT worker', motivated by concerns for a more fulfilling life beyond work and not to miss out on children growing up.[5] Indeed, these lifestyle choices amongst male engineers were also identified by their female colleagues, many of whom identified 'an awful lot more guys taking lifestyle choices around their career: going to companies where they'll be able to leave at half four to get home and actually have dinner with their kids. Whereas in the tech sector that was unheard before' (Marketing Manager, female, US IT MNC, Dublin). Similar observations were also identified at the HR level:

'When we offered the opportunity to purchase days [off work], 266 people on this site bought extra vacation days and 60 percent of them were guys. ... Men have greater expectations now and they're a lot more involved in the house and with their children.' Human Resources Manager, female, IT MNC, Dublin

Encouragingly, these experiences begin to challenge the common portrayal of men as merely 'shadowy figures' (Pocock et al. 2008: 26) within work-life balance debates, and the activities of some IT firms to support male employees who are 'going against the grain' (Ranson 2010) in assuming a greater proportion of childcare.[6]

Third, participants also identified significant work-life conflict as emergent from tensions between meeting shifting societal expectations around 'active fatherhood' (e.g. through reduced hours of paid work) and peer pressure from single male co-workers and managers without children (maintaining facetime at work); or else from male co-workers and managers who are able to devote long hours to work because they are supported by female partners who do not, and who fulfil the majority burden of childcare. These types of co-worker and managerial pressures were commonly articulated in relation to work team cultures of presenteeism:

> 'I would have thought, given that it's an IT company, that it would have been much more open to alternative work practices but actually there is still a culture of presenteeism. We even had one example today where one guy talked about their manager saying something like, 'I just don't buy into this work from home stuff, I want to come out and see you at your desk'. It's a tough sell to managers in particular.'
> Human Resources Manager, female, IT MNC, Dublin

Such peer pressure effects are particularly powerful in the computer software industry as a function of decentralised management structures in which autonomous work teams are disciplined by subtle yet strong systems of 'subjective' or 'normative' control through peer pressure (Perlow 1997; Kunda 2006) and inculcation of a professional/self-management ethos in which workers internalise a drive for long hours. In short, being 'fully engaged at work means under-involved elsewhere' (DTI 2004: 16). Participants commonly articulated 'worries about not letting the team down' and constraints on securing promotion, based on their inability (and unwillingness) to convey a depth of commitment to IT work through presenteeism and work addiction. These tensions were reconciled in different ways, also pointing to a gulf between some men's stated aspirations with respect to fathering/caring roles and their observed behaviour in practice:

> 'You try and make the work-life balance equitable between home and family … whereas if you ask my wife she'd say, "No, you're probably never late to a business meeting but you're often late home for dinner", so I still haven't quite got that balance right yet.' CEO, father, four children, indigenous IT SME, Dublin

On one level, these examples are certainly not inconsistent with gendered asymmetries of work-life reconciliation evident in previous studies, which suggest that the time men spend in paid employment determines what time they have left for caring (whereas the time women spend caring for their families determines how much time they have for paid employment) (Hewitt 1993: 4). At the same time, however, these work-life experiences of male IT workers suggest that the central character of the radically autonomous 'knowledge-driven agent' that has long populated regional learning accounts, who is always able and willing to put work first and to get the job done no matter what, is left wanting.

Back to the Future II: Silicon Cowgirls and Challenges of Girl Geek Motherhood

Earlier research has identified the distinctive occupational culture within information technology and software engineering as one 'requiring aggressive displays of technical self-confidence and hands-on ability for success, defining professional competence in hegemonically masculine terms and devaluing the gender characteristics of women' (Wright 1996: 86; see also Faulkner 2000; Wilson 2003; Crump et al. 2007). This occupational culture has historically been identified as discouraging women from entering STEM fields. More recently, however, women have begun to make significant inroads into the high-tech workforce: women now comprise an average 32.6% of high-tech workers in Europe (EU Labour Force Survey 2014), and over 1 million high-tech workers – plus owners of 173,000 high-tech firms – in the USA (US Census Bureau 2012). Underpinning this nascent feminisation of the IT sector have been the interventionist activities of a range of organisations – including Tech Divas, Girl Geek Dinners, Webgrrls, Women In Technology, Grace Hopper Celebration of Women in Computing (Anita Borg Institute) and the more recently formed Stemettes – which are actively seeking to disrupt popular portrayals of the masculine cultures of computing and engineering; and to increase the labour market advancement of women in multiple high-technology fields, by intervening at different points in the lifecourse (including schools, colleges, universities and industry). Nevertheless, work-life balance as a means for female recruitment and retention remains a contentious issue within the IT sector, encapsulated most recently in high-profile 2014 announcements by Apple and Facebook to offer funding to enable their female employees to freeze their eggs (Tran 2014).[7] The suggestion is that female technologists in their 20s and 30s should simply postpone the conflict between family and work in a manner that prioritises work and displaces conflict into the future (a WLB temporal fix?) rather than offering solutions to reconcile these tensions in the here and now. The message, then, is clear: 'work first, life later'.

Within the Dublin and Cambridge IT firm sample (N = 150), firms employed an average 19.3% women in their workforces locally, or a total of 1,551 women. On one level, the very presence of these female IT professionals exposes the false dualism within the regional learning agenda between a masculinist sphere of transcendence (science makes breakthroughs, is involved in change, in progress) occupied exclusively by men and a feminist sphere of immanence (reproduction, home) (Massey 1995: 189–190). Indeed, several of these women self-identified as 'workaholic programmers' and 'girl geeks' who enjoy their work and regularly 'just choose to work all the hours ... go home and often carry on – a few of us girls are as guilty as the boys for that too' (Principal IT Consultant, female, IT SME, UK SE region).

At the same time, however, the data also indicate how everyday struggles to juggle competing activities of work, home and family are particularly acute for female IT workers with young children. Consistent with previous research (see, e.g., O'Brien 2005; Lewis and Campbell 2007), despite a gender convergence in parents' contribution to childcare time, many of these women continued to undertake the majority share of household labour and childcare responsibilities.[8] As illustrated in Table 5.1, interviews pointed to the challenges of juggling temporally variable work demands with family life, juggling customer support and majority responsibility for childcare, and juggling senior management roles with motherhood and marriage. Within this framework, school holidays were identified as a particular pinch point by women in both the Irish and UK contexts. In Ireland, schools are required to be open 183 days per year (the UK has 190 statutory school days), which means that parents have to manage their children's care at least 182 days per year. And even for those parents not working weekends, this includes 130 weekdays when their children are not at school (and which is only partly covered by parents' annual vacation leave). Additional problems arise

Table 5.1   Lived experiences of work-family conflict: working mothers in the IT sector.

Work-family conflict examples	Indicative quotes from research interviews
Juggling temporally variable workloads with family life	'People who are leading sales revenue in IT essentially have no life for the last month of any quarter and absolutely no life for the last six weeks of the financial year. So from mid-October right through to December we could be doing anywhere between 40 and 60% of our revenues. You try and marry that with family life, with anything else going on outside work, and it quickly becomes a difficulty.' Software Business Development Manager, female, two young children, 3-day work week, MNC, Dublin
Juggling customer support and majority childcare responsibility	'We've got really large companies [as customers], they're demanding and expect top-notch service, and whenever there's a problem they'll ring, say quarter past six, that's happened to me loads of times – you can't say "sorry, I'm just in the middle of cooking a meal here" or "reading a bedtime story". Basically tough, you just have to get on with it.' Software Business Development Manager, female, two young children, 3-day work week, MNC, Dublin
Juggling senior management roles with motherhood and marriage	'I'm the CEO of [IT company] and I'm also the mum of two kids. So as an employer, I'm thinking about how I want to create an environment for the people who work in my company, and also honour the commitments I have to my shareholders … and also wanting to be successful as a mother. Just coming here tonight, I had a squabble with my husband – it's his birthday. I had a birthday celebration set up for him, which I told him had to take exactly 20 minutes. Anyway, it was not a happy moment, and that caused marital strain, let's say.' CEO, female, IT start-up, UK SE region

for working mothers whose young children are just starting school, which requires that parents also attend for part of the day.

The interview data also suggest a variability in the intensity of work-life conflict experienced by female IT professionals over the lifecourse, with many younger single women able to devote themselves to long hours at work, in contrast to female returners who experience the most acute forms of work-life conflict. This during 'the crucial period immediately after becoming a mother, when childcare becomes necessary and when women face loss of seniority, occupational status, lifetime earnings and pensions if they take extended breaks from employment' (Dex and Joshi 1999: 642):

> 'I was working for [large IT firm], and in my last year I had my son. I had 300 people working for me: you think to yourself, "I eat nails for breakfast, I'm gonna have a child and I'll be right back in there, grrrr". And the reality is, it's not that way, because all of a sudden you have something that you actually care more deeply about than your job.' CEO, female, two children, MNC, UK SE region

Female participants also explored the longer-term implications of these conflicts, and the need for assistance beyond simply 'muddling through'. The following extended interview excerpt is particularly instructive and far from unique amongst my research participant sample:

> 'Your life is lived at a pace, it's lived at a speed, if you're working, both of you all week and you're constantly running to either get to work, to get home from work, to get a meal on, to get the laundry done, homework – you actually don't know what it is to relax. Just keeping up with the kids' schedule, you actually can construct that kind of a momentum for yourself during your free time in a way that you don't even realise. You try and do the jigsaw on the floor while doing the dinner, while checking to see that the laundry hasn't yet finished, while trying to maybe hold a conversation with my other daughter, just juggling all these balls at the same time. So all this constant running, the only answer is just to have more time. Spontaneity goes out the window, that's the real pity of it, particularly as children get older, then they have their own schedules as well and everything has to be planned at that level of detail otherwise you just can't do it all, it's very exhausting. From my kids' point of view, I became much more aware of what I had missed out the first time round: any of the days off I had with my first child I very often ended up doing something work related. With my second child, it was an opportunity to reflect and say, well is this the way you want family life to be? Is this what's important, is this what's going to ultimately be the foundation of your children's memories and their success in life, and I just thought it wasn't, you know. If you just try and deal with these things on solving the problem, which we're all kind of quite good at doing, yeah, then you'll solve today's problem and you might plan a bit ahead for tomorrow. But the only way I could kind of make a decision about what was right for us as a family was to play forward 20 years and say, so in 20 years' time what will we all have remembered of this period in our lives? What will the children remember? What will their experience of being a

child in our home be?What will my personal experience be?What will [my husband]'s experience be? And I just figured there will be a lot more money in the bank, that's if we're still alive, if we're all still talking to each other, if, you know, the kids haven't gone off the rails because we haven't had time to sit down and talk because there hasn't been proper family conversation. So I approached [current employer] and initially they refused me parental leave!' Software Business Development Manager, female, two young children, 3-day work week, MNC, Dublin

Aside from highlighting the wider consequences of employers externalising the social costs of IT work onto workers and their families, this powerful quote also demonstrates how multiple pressures of work-life conflict combine, with clear negative work-to-home spillovers – and hence a clear need for WLB provision by employers. It also provides a clear illustration of how women often tend to experience conflicts between work, home and family *simultaneously* – this contrasted with previous work which suggests that men enact their multiple work-life roles *sequentially* (Sirianni and Negrey 2000). Previous research has also suggested that for many women the roles of work and home have effectively been reversed: home as the place in which there is too much (unpaid work) to do in too little time, and the paid workplace in contrast as offering stimulation, guidance, companionship and belonging (see Hochschild 1997). However, this neat binary reversal was not commonly articulated amongst my own participant sample.

A particularly useful framework for understanding the gendering of work-life conflicts is provided by Hardhill and van Loon (2006). They argue that the 'identity risks' which emerge from the tensions of pursuing parenthood with a professional job place particular pressures on women because the identity of being 'a good mother' invokes an everyday presence and involvement in childrearing that is absent from dominant societal expectations of what constitutes 'a good father'.[9] Indeed, the most extreme instances of work-life conflict documented in this research were also articulated by working mothers. This included one Software Business Development Manager at a US-owned multinational in Dublin 'sitting at our offices in town, seven and a half months pregnant, two in the morning, negotiating deals'. And another female CEO at an indigenous IT company who, following the birth of her son, 'went back to work within 4 weeks and brought him in with me – and to some meetings – and I just told people it was part of working/running an IT company and having a life'. In Cambridge, such conflicts even began in the delivery room:

'When I was having my contractions I was making copious notes for my deputy thinking, "Oh, I'm going to be out for a while now". Then when I was on maternity leave, I'd be breastfeeding and I'd be answering phone calls, I tended to carry on working through that. That's why I cut short my maternity leave to six weeks when I could have had three months off. I thought "I'm working from home anyway, I might as well come back to work". I left home when my children were a couple of months old, I did it with both of them, so yeah, I missed out a lot.' Director of IT, MNC, UK SE region

This framework usefully allows for a more nuanced conceptualisation of gender that avoids falling back repeatedly on gender as only referring to differing positions of male and female employees, instead considering workers as embedded in multiple gendering and sexing practices, involving employers, colleagues, spouses, other family members and the welfare state (following McDowell 1997; England and Lawson 2005; Smithson and Stokoe 2005).[10] As such, negative experiences of work-life conflict are not simply a narrow productionist function of corporate decisions and work pressure, but also emerge from deeply rooted ideas, scripts and codes about marriage and gender roles carried by both men and women, and continually invoked as they negotiate relationships across the work–home boundary with family members, bosses, co-workers and customers.

Sources of Work-Life Conflict Beyond Working Parents?

'It does still seem that the *only* possible reason anyone could be asking for this is because they're a new parent. That if you're not a young parent, then get to work! Why wouldn't you just want to be here at work all of your life?! There isn't any expectation that other people with other circumstances might need the same. It creates potential conflict within the workplace.' Principal IT Consultant, female, IT SME, UK SE region

The data suggest that experiences of work-life conflict in the Irish and UK IT sector are divergent from heroic portrayals of Silicon Cowboys, posing particular challenges for working parents. However, it is also recognised that persistent gender inequalities continue to be reinforced by dangerously divisive policy and workplace discourses which position WLB as somehow the exclusive preserve of working mothers (see Lewis et al. 2003; Emslie and Hunt 2009). In addition to domestic labour and childcare responsibilities (including the school run, relieving the nanny, attendance at school sports events, parent–teacher meetings), IT professionals with and without children identified a wide range of other personal responsibilities, commitments and interests outside of their paid work. Examples included home self-build, Cub Scout group leadership, care for elder parents, choral singing, acting, voluntary work, charity fundraising, international travel, home schooling children, horse grooming, writing novels, swimming training, part-time postgraduate study and labour organising. Add to these the kinds of daily work demands placed on IT workers as a function of time-based competition, and the everyday stresses and strains of juggling work, home and personal life in the Irish and UK IT sector beyond working parents become clearer:

'To me it's not only a women's issue. Everyone has work-life issues, they've all sorts of things they want to do, they want to play football, they want to go to the pub, commuting is a problem for everybody, it's everybody trying to cope with the pressure of life. So anything we can do to help people cope, makes a happy employee.' Director of Development, female, IT MNC, Dublin

In addition, research participants also pointed to the merging of work and personal life that results amongst project-based (portfolio) workers and self-employed IT professionals, who are largely responsible for their own work effort. Thus, while it is commonly accepted within the WLB literature that widespread conflicts between work, home and family are often simultaneously externally imposed by employers (e.g. increased work hours, unpredictable hours, downsizing, restructuring and an externalisation of the social costs of sustaining knowledge work) and internally imposed by workers themselves (e.g. self-blame, professional identities rooted in perpetual busyness as a badge of honour, career ambition), quite different work-life experiences are evident amongst less secure, less well-paid, project-based contractors. In particular, some participants highlighted experiences of overwork amongst project-based contractors keen 'to make hay while the sun shines'. Likewise amongst self-employed workers in my study for whom the categories of employer and employee are blurred, and many of whom continue to work long hours despite an original concern in becoming their own boss to increase their work-life autonomy relative to previous employers.

Importantly, other studies have also identified work-life conflicts amongst project-based workers as emergent from spending some (and in some cases most) of their work week away from home at a client site (Ahuja et al. 2007), and/or who are employed via third-party labour market intermediaries, which reduces their bargaining position, autonomy and choice in the paid work domain (e.g. De Bruin and Dupuis 2004; Doussard 2013).

Painful Outcomes of Work-Life Conflict in IT (on the *Need for Work-Life Provision*)

'In the IT industry, there's a particular attitude … I call the "jelly bean jar syndrome". That's because you have a jelly bean jar on your desk, or there's a pool table or there's bean bags in the corner and you don't have to wear a shirt and tie, there's this belief created that you're part of this revolutionary type organisation. But what they don't realise is that they're being exploited.' National Officer, Amicus Ireland

Emergent from the multiple sources of work-life conflict identified above, the most common outcomes – or negative work-to-home spillovers – identified here amongst IT professionals in Ireland and the UK include missing out on children's activities; interrupted sleep patterns; stress and exhaustion impacting on relationships with children and partners (e.g. impatience with children, time with partner curtailed, reduced frequency of family meals); working when feeling unwell; missing out on leisure time and hobbies; and 'taking work on holiday with me, and coming back from holiday and it feeling like I didn't have one'. At the broadest level (Figure 5.1a), these spillovers are evident in the aggregate IT worker survey data (N = 162), and are particularly evident amongst working parents of whom

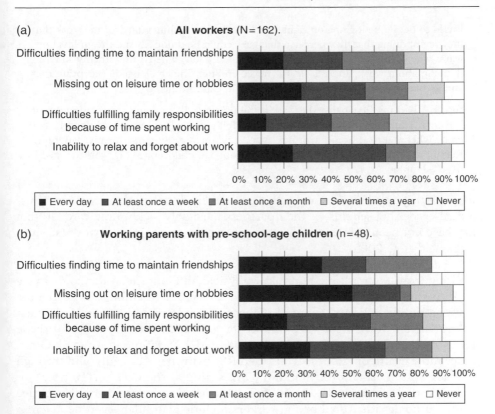

Figure 5.1   Lived experiences of work-life conflict: IT worker survey (2008).

the majority experienced the following work-life conflicts on a weekly or daily frequency: difficulties finding time to maintain friendships (53%), inability to relax and forget about work (69%), difficulties fulfilling family responsibilities because of work (53%), and missing out on leisure time or hobbies (68%). Moreover, these figures are typically higher for working parents with pre-school-age children (Figure 5.1b), with the greatest challenges evident amongst women with dependent children and a partner working full-time outside the home.[11] Similarly, the in-depth interviews also point to significant concerns amongst workers around personal 'care deficits', rooted in the perceived negative impacts of IT work demands on children's well-being, including constraints on workers' abilities to be home regularly for meal times, bath times and bedtime stories, to attend children's school sports events or assemblies, and associated feelings of guilt:

'The women I work with, especially who've got families, are guilty all the time. So when they get home they're thinking about the team that they left who are still working on a project, and they feel like, "God, you know, if I was a real go-getter like

I used to be, I'd be logging on right now". They start feeling guilty for the work that they're not doing that they perceive that they could be doing ... And the guilt is both ways: when I'm at work I'm thinking about my kid, when I'm with my kid I'm thinking about work, and it's almost never-ending. The challenge is accepting that and getting over it'. Women's IT career coach, Cambridge

'When I had my first baby, she was in the crèche full time and it broke my heart leaving her in and I felt guilty all the time. I felt guilty when I was at work and I felt guilty when I was at home with her. Also, it was a very long day for my daughter.' Recruitment Manager, female, two pre-school-age children, 3-day work week, IT MNC, Dublin

Some participants also explored these care deficits in relation to crossover stress in dual-earner households, as their partners inevitably pick up the care tab that they have not been able to pick up through over-commitments at work (see also McDonald et al. 2005).

In addition, workers also identified a series of negative health outcomes as resulting from the multiple temporal, physical and emotional demands of IT work. As summarised by one female software designer in Dublin, these include 'working public holidays, some of the engineers do that. Working even when feeling unwell, yes. Not having time to take a lunch break, yes. All of those things are true. I haven't exactly been exhausted, but definitely stressed. Other undesirable health outcomes? Really bad RSI in my hand actually, and through my back and my shoulder. You know, I'm sitting at the desk for 10 hours or whatever, and that hurts like hell. My health certainly suffers.' Other negative health outcomes identified by IT workers include glandular fever, headaches, high blood pressure, muscle aches, irregular sleep patterns and stress, with the most extreme outcome of long hours working in IT identified in the Dublin context:

'The whole team worked huge hours. In fact, my now brother-in-law who also worked at [previous IT company] had a car crash because he'd just worked a 36-hour shift and then drove home and of course just fell asleep at the wheel and BANG! – wrote off three cars and did a lot of damage. So I said to their boss, "Look, this can't happen again", but even then people *still* continue to work a lot of hours.' Software Developer, father of two pre-school-age children, IT MNC, Dublin

Significantly, these identified negative health outcomes of work-life conflict among IT professionals in Ireland and the UK are consistent with other studies documenting similar problems of stress and exhaustion amongst IT workers (e.g. Fujigaki and Mori 1997; Fujigaki et al. 1994; Moore 2000; Darshan et al. 2013).

Far from simply problematic for workers and their families, participants also identified the wider implications of excessive work-life conflict for the competitiveness of their respective firms. This was most commonly framed in terms of highly qualified female IT professionals with young children taking compromise jobs which reduce work strain but ultimately under-utilise their skill-sets, knowledge and experience:

> 'When you come back to work and you've got a 12-month-old baby, teething, got a cold, whatever, they're up two or three times in the night, you go to work the next day, you're tired, you're not necessarily thinking as properly as you should be. So then to come into a less responsible job, it's not as challenging and it's not as motivating but that's fine. Because I'm not doing this to further my career, I'm doing this as something different to do to keep me working until the children are old enough where I can go back and do something that I really want to do ... that the children are that bit older that you're not getting disturbed sleep and then it will all pull together. That's my plan.' Sales Manager (previously engineer), female, two pre-school children, 2-day work week, ICT MNC, Cambridge

This quote points to the ongoing need for better support in the IT sector for female returners after maternity leave (see also Panteli and Pen 2007). For Rutherford (2001), time becomes a patriarchal resource, in which long hours work cultures in the IT sector act as a means of social closure to exclude female technologists with young families. Or as one female programmer put it: 'Look at the productivity you lose after women have children if you're not supporting them.' As such, the business case for WLB here is not just about the costs of replacing female employees who have left the company, but also about the limits to learning resulting from the loss of workers as embodied expertise, skills and knowledge:

> 'Right across the board they have really cut back on part-time working and I think that's to the detriment of the organisation because they've lost a lot of really good strong people over the years – people like myself who've had a lot to bear. Because the norm is that people just get out, they leave when they find the situation untenable at home, then they end up leaving and either consulting or contracting.' Software Business Development Manager, female, two young children, 3-day work week, MNC, Dublin

Beyond individual IT firms, long hours, negative work-to-home spillovers and a lack of effective WLB provision and support were also identified as a major factor in some women quitting the IT sector altogether, resulting in an industrial loss of embodied skills, training investment and accumulated experience. Similar motivations have also been documented amongst surveys of 716 female workers quitting high-tech jobs in the USA (Snyder 2014) and of 3,254 mothers in the UK extending beyond the high-tech sector (EHRC/BIS 2015).[12]

Work-Life Conflict, in What Sense a Regional Problem?

The multiple outcomes of work-life conflict for workers, families, firms and industry as described above must also be understood as geographically uneven. Research participants in Dublin were consistently much more vocal about their regular experiences of work-life conflict than were their IT colleagues in Cambridge. Likewise the IT worker survey (N = 162, 2008) documented a greater prevalence of daily/weekly work-life conflict amongst IT workers in the Dublin cohort than the Cambridge cohort across *all four* types of negative work-to-home spillover: (i) inabilities to relax and forget about work (77% cf. 60%); (ii) difficulties fulfilling family responsibilities because of time spent working (46% cf. 39%); (iii) missing out on leisure time or hobbies (64% cf. 54%); and (iv) difficulties finding time to maintain friendships (53% cf. 44%). These differences are informed by different gendered welfare regimes and work-life legislative provision in Ireland and the UK in relation to paid work and care (see Table 4.1). Crucially, there is no legal right to work part-time in Ireland – part-time work, job sharing, flextime and teleworking are all at the discretion of employers (Russell et al. 2009). In addition, statutory paid maternity leave entitlement is lower in Ireland than in the UK (26 weeks plus 16 weeks unpaid compared with statutory 39 weeks paid, plus 13 weeks unpaid in the UK). Similarly, there is no statutory provision for paternity leave in Ireland (compared with the UK at 1 or 2 weeks at 90% pay). Reinforcing these challenges, childcare costs are higher in Ireland than in the UK, costing 47% of Ireland's average wage for two children in a single-earner household, compared with a figure of 10% for the UK (figures calculated on the basis of one month of full-time care without allowance for government subsidies, average wage as defined by the OECD 2007a *Babies and Bosses* report).

Moreover, in a major EU comparative study on *The Rationale of Motherhood Choices* (European Commission 2006), Ireland was ranked last in terms of public and publicly funded childcare systems.[13] And in terms of cost, Irish parents have been identified as paying on average 20% of their annual income towards childcare, almost double the European average of 12% (Forum on the Workplace of the Future 2005: 62). In other words, Irish working parents bear a greater burden to meet their childcare needs in comparison with their peers across the EU (Redmond et al. 2006). Against this backdrop, employer-provided extra-statutory maternity leave, paternity leave, flexible working arrangements and childcare options have an enhanced meaning and significance in Dublin compared to Cambridge.

Higher levels of work-life conflict in Dublin must also be understood in relation to Dublin's booming housing market, which has seen strong price increases over the last two decades as a distinctive element of the Celtic Tiger growth phenomenon (Fahey and Duffy 2008).[14] The response by many IT workers has been that 'people now have chosen to move out of the city to some of the outlying counties around Dublin, so their commute can take anywhere up to an hour, an hour and a half. So it's not just work, it's the getting *to and from* work which is taking out of your personal time and your ability to manage your personal life' (Human Resources Manager, female, US-owned IT MNC, Dublin). Participants in my study had moved out of Dublin city to surrounding towns and villages including Drogheda, Bray, Swords, Navan, Kildare, Newbridge and Skerries (see Figure 5.2), increasing the length of their daily commutes in the process (and, for

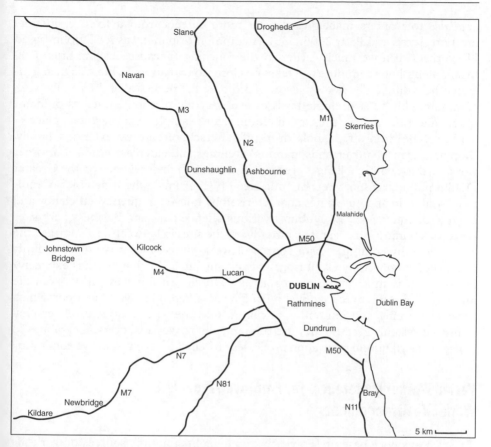

Figure 5.2   Dublin and its commuter environs.

many, the associated costs of extended childcare as already identified above).
Likewise perceived negative outcomes for workers' productivity and effective use
of time:

> 'When the commute gets to an hour and a half, an hour and three quarters [each
> way], you've got to ask yourself why am I doing this? What's the point? What can I
> be doing with that three hours of my day?' Senior Software Developer, male, two
> children (school age), now working from home full-time, IT MNC, Dublin

These expanding travel-to-work areas and associated patterns of traffic congestion
must therefore be understood as key features of Dublin as an (increasingly
sprawling) 'urban carescape' (McKie et al. 2002), in which the geographical distribution and spacing of jobs, housing and services within a particular urban area
also determine the working time arrangements and childcare options *actually*

available to workers managing one or more careers from a fixed residential location (Jarvis and Pratt 2006: 333). A common response was for Dublin-based IT workers to leave much earlier in the morning (because 'leaving after 7:00 means a one-hour commute will take me closer to two hours!'), but this then generated new difficulties as they try to coordinate the necessity of their daily work commute with other multiple tasks of everyday life distributed across different spaces and times within the city (including, for example, their partner's place of work, crèche drop-off, schools drop-off, afterschool care by extended family, shopping, gyms, bank errands, parents' evenings and eldercare visits). For other workers, these daily challenges are amplified through their reliance on the Greater Dublin public transport network and need to coordinate the timetables of multiple local train and bus routes characterised by uneven frequency, efficiency and degrees of connectivity to Dublin's various science park and business park sites (and back home again, via many other work-life sites in between). In contrast, the role of lengthy commutes as a source of work-life conflict was not commonly articulated amongst my Cambridge participants, in a city that also has expensive housing but is much more compact than Dublin. In this way, these work-life infrastructures are more than just a benign social reproductive stage upon which regional learning and innovation is played out, they are *constitutive of* regional industrial systems because they are fundamental to workers' everyday abilities to engage in learning and innovation processes on behalf of their respective employers.

Varied Worker Preferences for Employer-Provided Work-Life Arrangements

The rich evidence base presented above highlights multiple sources of work-life conflict amongst IT workers as a function of multiple work demands juggled with a range of personal responsibilities and interests outside work, giving rise to a series of negative outcomes at the level of workers, families and firms. In response to these everyday sources of work-life conflict and resultant negative work-to-home spillovers, research participants were asked to identify their preferred employer-provided policies and practices which are most effective in helping them to reconcile everyday work-life conflicts. Strikingly, despite a clear gendering of work-life conflict as described, the results do not indicate a similarly clear gendering of preferred WLB arrangements. Rather, the interviews suggest that there is no panacea: WLB requirements vary within gender groups (by job function, department, household situation) and for individual workers over time. Thus, for example, while working mothers and fathers with young families often identified extra-statutory provision of maternity, paternity and parental leave as their preferred WLB arrangement, this was less important for working parents with children of school age, and an irrelevance for research participants without children. The point is that different home and work factors mean work-life

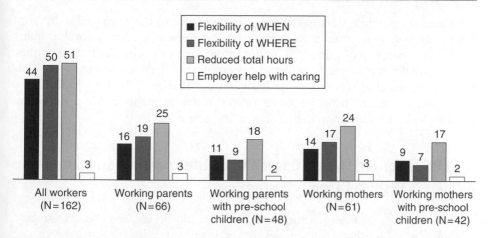

Figure 5.3 What are workers' preferred types of employer-provided work-life arrangements? IT worker survey (N = 162, 2008).

provisions suitable for one class of employees at one point in time may have little or no effect in reducing work-life conflict for another class (see also Phillips et al. 2002). Similarly within the IT worker survey, workers identified a range of preferences for different types of work-life provision by employers, and these are shown in Figure 5.3. Again there is no single, clear-cut, magic bullet favorite.

Further reinforcing this complexity, the kinds of WLB provisions required by a particular individual at one point in the lifecourse are also changeable over time in relation to major life events, including labour market entry, marriage, divorce, childrearing, moving house, nest-leaving and eldercare responsibilities (see also Charles and Harris 2007; Bowlby et al. 2010).

Ultimately, this diversity points to the urgent need for more comprehensive suites/bundles of employer-provided work-life arrangements to respond to variations in WLB requirements across workers as a function of occupational role, household situation, caring responsibilities and personal life interests. Research participants also highlighted common disjunctures within their respective firms between the kinds of work-life arrangements they favoured most (particularly reduced hours and flexplace), and the kinds of arrangements most commonly available to them (flextime):

'So with [previous company] about three or four years ago, they introduced flex time benefits, so you can buy extra holiday if you want, take more leave. But that's quite an easy thing to implement, it doesn't change the underlying long hours work culture that might be causing some of the WLB issues that bubble up. A lot of these are interim measures, they are just treating symptoms, they're not really getting to the root cause of the problem, they're just sticking plasters over the dam to try and keep people in the workplace.' Principal IT Consultant, female, IT SME (previously at IT MNC), UK SE region

Fundamentally, while flextime allows workers to rejig the temporal pattern of hours worked, it is problematic because it merely addresses what for many workers is a *symptom* of work-life conflict rather than its deeper underlying *causes*: namely, total hours worked and lengthy commutes. These criticisms were commonly articulated by workers in the Irish context. But how, then, do these worker preferences compare with the kinds of WLB arrangements made available in practice by IT employers in the Irish and UK contexts?

Fruit Bowls for 'Girl Friday': Uneven Work-Life Provision, Stubborn Barriers to Positive Change

'They call me Girl Friday because I do Tuesdays and Fridays and then, "Oh is it Friday again, cor that's gone quick?" or, "[Name]'s in today" or, "It must be a Friday"'. Sales Manager (previously Software Engineer), female, two pre-school children, 2-day work week, ICT MNC, Cambridge

As shown in Table 5.2 the most commonly available types of employer-provided work-life arrangements amongst the IT employers surveyed are flextime (67% of firms), working from home one or two days a week (66% of firms) and part-time working (54% of firms).[15] Compare this with the much greater diversity of workers' WLB preferences identified earlier in this chapter which include, for example, the preferences of working mothers and fathers with young families for extra-statutory provision of maternity leave (available in only 22% of firms), extra-statutory paternity leave (available in only 11% of firms) and employer-subsidised childcare (available in only 6% of firms). Likewise identified worker preferences for compressed working weeks (available in only one third of firms) or term-time working (available in only one tenth of firms). Although within the context of prevailing gender relations which position women as primary caregivers, and an ongoing gender pay gap rooted in stubborn gender labour market inequalities of power and influence, women's 'preferences' for particular forms of (e.g. part-time) flexible working are never truly 'voluntary' (Rubery 1997; Figart and Mutari 1998).

The employer survey data also document unevenness in the *total suites of WLB provision* available to workers across different IT employers, in which the majority of firms provide less than 5 of the 15 different types of WLB arrangements identified in Table 5.2. (Two exemplar firms offered all 15 WLB arrangements, one as part of a larger MNC with offices in Dublin, and the other as a progressive Irish indigenous SME.)

Reinforcing this unevenness of work-life provision, particular WLB arrangements were even exposed to criticism and ridicule, such as one IT firm's provision of fruit bowls on the front desk in reception during their annual WLB Week which – in the context of lengthy commutes, excessive workloads, long work

Table 5.2   Employer provision of (formal) WLB arrangements, Dublin and Cambridge IT firm sample (2008).

Category	Formal WLB arrangement	Dublin N = 74	Cambridge N = 76	Combined N = 150
		%	%	%
Flexible work arrangements	**Flextime** (flexible beginning/end time, sometimes with core hours)	**73**	**61**	**67**
	Flexplace (work from home 1 or 2 days a week)	**74**	**58**	**66**
	Flexplace (work from home 3 or 4 days a week)	35	53	44
	Job sharing (one job undertaken by 2 or more persons)	9	4	7
	Annualised hours	8	11	9
Reduced work hours	**Part-time work**	**49**	**59**	**54**
	Compressed work weeks (total work hours in 4 days rather than 5)	31	29	37
	Term-time working	8	9	9
Personal leave	Extra-statutory maternity leave	32	12	22
	Extra-statutory paternity leave	14	9	11
	Career break/sabbatical	19	7	13
Practical help with childcare	Employer-subsidised childcare	4	8	6
	Information referral service for childcare	4	3	3
	Workplace nursery	3	1	2
Other	WLB counselling/training	15	4	9

hours, performance targets and peer pressure from work team colleagues – kind of misses the point. Nor is this an isolated case:

'I did some work at one firm and every time I went to the offices, they had these elaborate Zen rooms where you could go and download music, they had beanbags, all that kind of stuff. In all the times I went and visited, I never saw anybody using that. And I just thought, this is the red herring: who wants to be the one *seen* sleeping, if nobody else is doing it? And frankly, when I hear people talk about what would make a difference to them, it is not a Zen room! It's reduced hours.' Women's IT career coach, Cambridge

The pattern of WLB provision summarised in Table 5.2 also points to the need for more radical provision that also seeks to degender asymmetries of working time. Here, an analytical distinction is made between the prevalence of employer-provided work-life arrangements that *accept* the accommodation of working time to existing gendered divisions of labour (e.g. part-time working available in over half of the firms surveyed) and more radical provisions that *challenge* the models of male full-time employment/female primary carer by supporting the reallocation of childcare and reproductive tasks (e.g. extra-statutory paternity leave available in only 11% of firms surveyed, annualised hours (9%), WLB training (9%), employer-subsidised childcare (6%) or work-place nursery (2%); see Beechey and Perkins 1987: 107; Figart and Mutari 1998: 462).[16] As such, these IT employer data are consistent with a broader tendency for employers across multiple sectors to provide non-traditional working arrangements under the banner of work-life balance, but which run the risk of perpetuating uneven gender divisions of household reproductive labour, the downward occupational mobility of women returning from maternity leave, and/or long work hours.

In addition, the interviews identified a number of ongoing barriers to meaningful WLB provision, rooted in a limited recognition of the benefits of employer-provided work-life arrangements beyond individual workers and their families. These include limited managerial support, and perceptions within software teams that colleagues who avail themselves of non-traditional working arrangements (such as working from home, flextime or annualised hours) are letting the team down. Interestingly, some managers were critical whilst dealing with the same challenges themselves:

> 'As a parent I am very aware of the value of family life to me, as an individual, and therefore I wish to support fellow human beings in their enjoyment of that family life. It's just there has to come a point where you say, "Are you working for this business to move this business forward or is it a total convenience because it happens to fit very well with your family life?" And I think that's often quite a tricky one.'
> CEO and founder, female with two children, IT MNC, UK SE region

Other less nuanced examples included one female technologist in Cambridge who 'drops her daughter off at school, comes straight to the office, but gets yelled at because she's coming slightly late and going home slightly early. She takes lots home with her and does the work but she doesn't get any credit for it. They just assume that she's being lazy and she's slacking even though they know she's got kids' (Information Management Consultant, female, MNC, UK SE region). Similar negative experiences were also identified in Dublin:

> 'I find from a colleague's point of view, people who don't have kids, they don't understand. A couple of times if I was leaving at 4:30 somebody would say, "Oh, half day?" And I'd say, "Well no, I've been here since 8:00". I'm probably the most productive of anybody in the team!' Recruitment Manager, female, two pre-school-age children, 3-day work week, IT MNC, Dublin

Also emergent from ongoing scepticism amongst some managers of the wider business benefits of WLB provision beyond the individual workers who use them, research participants highlighted several instances where particular WLB arrangements provided by their IT employers were only available in practice to a select group of established senior colleagues. That is, those for whom the cost to the firm is deemed worthwhile. This included one software designer's immediate supervisor who 'works four days a week. She's essential to the company and so she went in to the MD. It would be a different case altogether if somebody who was less appreciated went in. So there *is* a practice but one of the more lower-down-the-ranks engineers wouldn't be able to' (software designer, female, Irish IT SME, Dublin). And again, these challenges were particularly apparent in smaller IT firms:

'A very important employee said, "I wouldn't mind getting summer off on a sabbatical". We said, "Absolutely no way, can't do it, we're too small". And he said, "Well I quit then", and then we said, "You can have a sabbatical! How long do you want?" [laughs]. We don't want to lose any employees, particularly the more experienced ones. If that's their life choice, are we going to lose them or are we going to have them under whatever *restrictions* they put on?' CEO, father, four children, Irish indigenous IT SME, Dublin

In the words of this CEO, then, WLB provision is seen as a 'restriction' on the company. Likewise other CEOs for whom work-life provision was identified as merely an additional cost to the firm, and variously described as 'prohibitive', 'burdensome', 'a significant challenge', 'causing work-flow problems' and a 'management headache' that 'needs to be reined back in'. And whilst these views were most commonly articulated by managers in smaller IT firms with limited resources, it is also important to note the views of one senior manager in a large US-owned IT multinational in Dublin for whom 'teams of young male developers are simply easier to manage'. Reinforcing these various constraints to extending WLB provision in the IT sector, participants identified their additional significance in the aftermath of recession.

'Struggling with the Juggling' in the Wake of Recession

The analysis presented in the main body of this chapter draws on data collected in Dublin and Cambridge between 2006 and 2008, just prior to the financial crisis and subsequent economic downturn. To begin to gauge the degree to which WLB concerns continued into that period, a second online survey of IT workers was undertaken November–December 2010 (N = 147), around two years into the Great Recession. These workers exhibit a similar skills profile to the pre-recession IT worker survey sample with an average 11.6 years' work experience in the IT sector across 3.8 different employers. Additionally, 39% had a Masters degree or higher; 70% were employed in technical managerial, technical or research roles at the time of the survey, and only 7% in HR, marketing or sales roles.

The recessionary worker survey data offer three insights. First, they evidence negative changes in many IT workers' qualitative experiences of work-life conflict through the economic downturn, as a function of 'more commuting to the office to be seen where previously you would have worked from home; doing more hours as you are required to if colleagues have left or you just want to look like a contributing member of the team; and a less predictable schedule as bosses try to come up with new urgent schemes to get us out of the current hole' (Senior Marketing Manager, female, IT MNC, UK SE region). Other drivers of negative change identified in the survey include 'reduced headcount and higher work-loads', less predictable schedules, more unpaid overtime, earlier start times, increased weekend working (so that 'rather than a 50-hour work week it is more like over 70 hours'), pay freezes, increased presenteeism, survivor shock, 'increased wage dependency on bonus incentives, which is also dependent on company profit', 'rates considerably lower than recruiters had been discussing with me in previous years', and 'working away from home more [which] has added to my overall hours, but none of travel time or accommodation costs are chargeable'. These negative changes were also identified as resulting in 'feelings of guilt at taking annual leave', despite 'my 4-year-old thinking it's normal for mummy to do conference calls at 7:30 am as I did this morning'. Likewise, workers identified 'cutbacks in activities without an obvious contribution to the bottom line, like knowledge-sharing communities'; and, as a function of office closures for cost savings, being 'forced to commute to London to work, adding 4 hours' travel a day (was 1 hour), so my days are much longer'. The overall result was usefully summed up by one participant: 'I like working, it would just be nice to occasionally have a break' (Technical Engineer, female, MNC, UK SE region). Indeed, only 1 of the 147 IT workers surveyed articulated a (qualified) improvement in their personal experience of work-life balance through the recessionary period, on the basis that:

> 'The company have been more amenable to discussions on more flexible arrange-ments. As a single mum I can work 3 days per week in school holidays and pretty much full-time in term time. They have cut expenditure and I have a better balance (in theory). Of course I still take on too much in holiday weeks and end up working more than contracted but the arrangement generally suits.' CRM Data and Digital Manager, female, IT MNC, UK SE region

These patterns are also evident in terms of three quarters of workers experiencing an inability to relax and forget about work on a daily or weekly basis in the recessionary period, compared with two thirds of workers in the earlier survey. Additionally, half of all workers also experience difficulties fulfilling family responsibilities because of work, difficulties finding time to maintain friendships, and miss out on leisure time or hobbies on a daily or weekly basis. And again, these challenges of work-life conflict are greatest amongst working parents, with the majority experiencing daily or weekly difficulties fulfilling family responsibilities because of time spent working. These problems were particularly apparent amongst working mothers on fractional contracts, in which workers often continued to maintain a near full-time workload, and made clear that 'it's so hard to train yourself away from being 100 percent constantly available' (Software Business Development Manager, female, two young children, 3-day work week, US-owned MNC, Dublin). Likewise insights from IT workers on fixed-term contracts, who now found themselves 'working longer hours in a naive attempt to "put myself above the rest"'. In one case, however, while 'the additional hours spent at work were unpaid, I felt I had to do this to stand me in better stead in getting an extension in my contract. This had a major effect on my work-life balance! Unfortunately my contract was not extended' (Network Coordinator/Technologist, female, IT SME, UK SE region).

Second, the data also document IT workers' experiences of shifting patterns of WLB provision by employers through the economic downturn. Most notably, while these fail to document any *widespread* formal rollback of work-life arrangements amongst IT employers, 40% of survey participants identified a reduced willingness *to use* formally provided WLB arrangements in practice compared with prior to the onset of recession – this typically premised on fears of a lack of informal ratification by employers and negative career outcomes. As such, the recession survey data suggest work-life conflict is even *more* salient for IT workers in the wake of the recession. While several employers surveyed were seeking to expand their WLB provision in order to improve talent recruitment and retention, other firms were reducing their work-life provision in pursuit of short-term cost savings. And in contrast to earlier recognition of the crucial role of managerial empathy in enabling WLB in practice, exclusionary cuts were also self-identified in female-founded and female-managed firms:

'I am the only woman in my company, with seven men working for me. Now if we had a woman apply for a job (which has not been the case up to now), I would as MD have serious concerns about maternity leave ..., it would just be too costly in the current climate to take the risk. It is awful to say as a woman but as an MD of a small (but growing) technology business, it is the reality on the ground. The government does not incentive us.' CEO, female, indigenous IT SME, Dublin

Moreover, in the Irish context, these recessionary challenges to work-life provision were reinforced at the national policy level by a reduction in the perceived significance of WLB as a key focus of the Social Partnership process agreed between the government, the Irish Business and Employers Confederation (IBEC) and the Irish Congress of Trades Unions (ICTU) – this relative to recessionary policy concerns around pensions, pay and job retention which were deemed much more pressing.

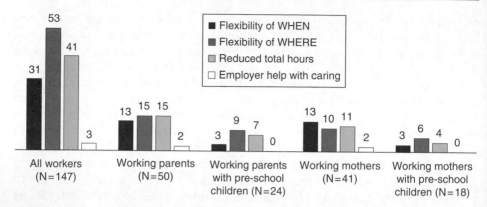

Figure 5.4   What are workers' preferred types of employer-provided work-life arrangements in the wake of recession? IT worker survey (N = 147, 2010).

Third, the data evidence recessionary shifts in workers' preferred employer-provided WLB arrangements (comparing Figures 5.3 and 5.4), with a lesser preference for reduced work hours (from almost a third of workers in the 2008 survey to only 5% of workers in 2010). This shift can be understood as a function of its associated reduction in monthly take-home pay, which is not feasible for many workers – especially in expensive property markets and with rising costs of living, which necessitate two incomes for households to get by. More generally, however, the recessionary data fail to identify any single 'magic bullet' worker preference, and hence a continuing need for varied suites of work-life provision by employers.

The core contradiction, then, is that IT workers are experiencing *greater* levels of work-life conflict – and hence *greater* need for assistance in reconciling competing activities of work, home and family – but this is coupled with a *reduced* willingness to use formally provided WLB arrangements in practice, for fear of lack of informal ratification by employers and negative career outcomes. This is reinforced by employer concerns around the costs of work-life provision. In short, the WLB business case is even more salient. In seeking to challenge this contradiction, and to integrate the mainstream focus of regional learning and innovation debates around (knowledge) production with concerns around social reproduction, family well-being and quality of life, the analysis presented here in Chapter 5 raises important questions. Specifically, what benefits do workers experience in practice (both at home and at work) from making use of the different kinds of WLB arrangements differently provided by IT employers? What, then, are the consequences of that WLB uptake for the kinds of social interaction and workplace activities that support learning and innovation within knowledge-intensive high-tech firms? And what are their consequences for shaping the kinds of cross-firm labour mobility pathways widely identified as a key mechanism for cross-firm knowledge transfer within dynamic regional industrial systems? These form the novel analytical focus of Chapters 6 and 7 respectively.

Notes

1 'By signing up for a project you agreed to do whatever was necessary for success. You agreed to forsake, if necessary, family, hobbies, friends … in effect declared, "I want to do this job and I'll give it my heart and soul"' (Kidder 1981: 63).

2 In contrast to this positive portrayal of programmers' intrinsic motivation and enjoyment of work, Perlow's (1997) account of the software engineering labour process connects long work hours to workers' concerns to be visible to their managers and to avoid negative assessments and appraisals.

3 ICT total workforce Ireland (2014): 88,485 (18,570 women); cf. UK 1,494,051 (257,777 women) (Eurostat 2015, ICT specialists in the labour force (isoc_sks) and ICT specialists by gender (isok_sks_itsps) databases, accessed 21 April 2016).

4 Similarly, the UK's Labour Force Survey shows that two thirds of the IT worker cohort (N = 1,457) work 35–48 hours per week, with 8% also working 10 or more hours of unpaid overtime (LFS 2014, NACE Sector J: Information and Communication Workers).

5 Similar trends have also been identified amongst some male engineers in Silicon Valley, where 'coming in early is the new staying late' amongst male engineers with young children (see Dembosky 2013).

6 While women still undertake the majority of unwaged carework, a growing body of research suggests some men – and especially fathers – are engaging more in the everyday tasks of social reproduction (e.g. Doucet 2006; Hook 2006; Rochlen et al. 2008). This is understood as a function of shifting cultural understandings about fathering, with increased expectations for fathers to participate in caregiving and other domestic tasks (Dermott and Miller 2015).

7 Facebook offers up to $20,000 (£12,500) to fund egg freezing, plus £300 for subsequent storage. This has been criticised by some commentators as 'breaking employees down to discrete biological functions, trying to control each one that we might work better, harder, more consistently' (Williams 2014). Apple's workforce is 70% male, compared with 69% for Facebook (Tran 2014).

8 For example, Ireland's National Time-Use Survey shows that women in Ireland spend five times longer on caring activities during the week than men (McGinnity et al. 2008: 205).

9 Case in point: 'So when the children are sick, it's normally me that has to take the time off, not my husband. And if I've got meetings and things set up then that is a problem. But then working part-time it means that generally if the child's only sick for like a week at a time you're only actually missing two days as opposed to having to take a whole week off.' Sales Manager (previously Software Engineer), female, two pre-school children, 2-day work week, US-owned ICT MNC, Cambridge.

10 Lone parenthood raises additional considerations but these were not articulated in any depth amongst my research participant sample. More widely, the experiences of single-parent mothers and fathers are less well documented in the expansive work-life literature than those of other family forms. But relevant studies include DeBord et al. (2000), Mason (2003) and Minnotte (2012).

11 The IT worker survey (N = 162) documented the majority of 'women with dependent children and a partner working full-time outside the home' experiencing all four

dimensions of work-life conflict on a daily or weekly basis: inability to relax and forget about work (77%); difficulties fulfilling family responsibilities because of work (64%); missing out on leisure time or hobbies (79%); and finding time to maintain friendships (72%).

12 Snyder (2014) surveyed 716 women who had quit the US high-tech sector. The main reasons cited include a lack of flexible work arrangements, an unsupportive work environment and a salary that was inadequate to pay for childcare; 625 of these women said they had no plans to return to high tech.

13 15 EU countries were ranked according to how supportive their childcare system is of the dual-earner model. Denmark came top with a score of 88.5 (out of 100), followed by Sweden (78.9). The average rating of support was 40.9. Ireland scored only 5.6 (European Commission 2006). To bridge this care gap amongst working parents, Ireland's Quarterly National Household Survey evidences the significant use of unpaid relatives by almost half of all households with primary school children.

14 Lift-off for the Irish housing market began in 1994, with nominal second-hand house prices having increased by 4.6 times (and new house prices by 3.7 times) between 1994 and 2005 (Fahey and Duffy 2008: 124). In Dublin, the average new house price for which loans were approved increased by 136% between 1994 and 1999 alone (Kirby 2002: 48). These house price increases can be understood as a function of population growth, rapidly rising real incomes, a reduction in income taxes and a fall in mortgage interest rates.

15 This dominance of flextime, working from home and part-time working as the three most commonly provided WLB arrangements is also evident in the UK's Fourth National Work-Life Balance Survey (2011), based on a much larger sample of 2,767 workers across multiple sectors. Similar patterns are also evident in Ireland's earlier national WLB survey *Off The Treadmill* (Drew et al. 2002) based on a national sample of 912 employers and 1,006 workers across multiple sectors.

16 Indeed, a more radical approach to work-life balance provision should also involve shifting the definition of the problem from (women) workers back onto organisations themselves.

Chapter Six
Overcoming Work-Life Conflict and the Gendered Limits to Learning and Innovation?

Introduction

'There has to be recognition there that people have lives outside work, that people have families, that people have children that need to be cared for, in terms of work-life balance. But on the ground I don't think enough is happening. We have to be brutally honest here: it's cost – if you look at the small firms they're terrified of this. Maybe the resistance is that people are not fully informed of the realities.' Programme Manager Equality and Policy, Irish Congress of Trade Unions, Dublin

The everyday challenges faced by workers struggling to juggle competing commitments of paid work, home and family remain stubbornly persistent and distinctly gendered. In response to commonly experienced work-life conflicts amongst information technology workers, Chapter 5 documented the kinds of employer-provided work-life arrangements that different workers find most useful in ameliorating those difficulties. In the absence of support, that analysis highlighted connections between poor work-life balance and increased stress, strained familial relationships and gendered labour market inequality. Exacerbating these problems, the 'global' economic downturn was also shown to have created new work-life demands amongst IT workers, through rapid and dramatic labour market change, heightened fears of job loss, increased workloads and understaffing. With employers keen to effect cost savings, their provision of work-life arrangements designed to help workers reconcile work, home and

Work-Life Advantage: Sustaining Regional Learning and Innovation, First Edition. Al James.
© 2018 John Wiley & Sons Ltd. Published 2018 by John Wiley & Sons Ltd.

family was also shown to have been rolled back in a series of subtle and less subtle ways. Accordingly, the 'business case' for WLB becomes even *more* salient.

The policy-espoused employer benefits of work-life provision find some empirical support in academic studies and include increased productivity, improved employee retention, improved recruitment, decreased absenteeism, decreased employee turnover and improved financial performance (see Chapter 3). Yet many employers still regard WLB provision as too costly, and 'the luxuries of a booming economy that cannot be sustained' (Leighton and Gregory 2011: 11). As such, work-life provision is narrowly positioned as merely an extra administrative cost to the firm, rather than as a viable mechanism for enhancing firm competitiveness in pursuit of economic growth. A major sticking point for change is the continued dearth of empirical evidence to support the so-called WLB 'business case' in practice. Likewise, 'few scholars have demonstrated the mechanisms through which such [WLB] policies function (or do not) to enhance firm performance' (Eaton 2003: 145–146). In response, this chapter explores the learning and innovation advantages that can result from work-life provision within knowledge-intensive firms, as part of an alternative mutual gains research agenda. (In Chapter 7 I expand this intra-firm focus to explore the connections between work-life balance and *cross-firm* learning networks.)

The first part of this chapter explores the benefits experienced by IT workers and their families from using differently preferred work-life arrangements to overcome commonly experienced conflicts between work, home and family (WLB dual agenda part I). Against this backdrop, the second part of the chapter explores the mechanisms through which employer-provided WLB arrangements can also enhance firms' everyday capacities for learning and innovation (WLB dual agenda part II). At the broadest level, almost two thirds of managers responding to the IT employer survey (N = 150) indicated improved worker productivity as a result of their total bundles of WLB provision and take-up, and over half 'an improved corporate environment for learning and creativity'. Importantly, these managerial perceptions are also consistent with measured improvements in firm performance over the same time period. Chapter 6 delineates three WLB learning mechanisms which are recognised by research participants to underpin these synergies, centred on: (i) WLB and increased self-determination of work; (ii) WLB and increased work team heterogeneity; and (iii) WLB and enhanced sustainability of learning. As such, this chapter highlights the dangerous irony of the rollback of WLB provision by employers in pursuit of short-term recessionary cost savings, at the expense of firms' longer-term innovative capacities.

The final part of Chapter 6 identifies two additional mechanisms – (iv) WLB backlash and constrained worker agency; and (v) WLB and limits on communication – through which some limits to firms' innovative capacities also emerge from workers' use of non-traditional working arrangements in pursuit of improved WLB. These two mechanisms of constraint are explored in relation to the face-to-face demands of agile software development, and the role of workplace interventions and contextual factors in mediating those effects and minimising their negative impacts. This discussion critically examines the use of homeworking technologies to reduce those constraints. However, for some workers these work-life/communication technologies are also shown to be a double-edged sword, blurring further the complex spatial and temporal boundaries between 'work' and 'home' and generating new work-life conflicts.

Improving the Work-Lives of IT Workers and their Families (Mutual Gains, part I)

As identified in Chapter 3, the mainstream WLB business case evaluation research literature predominantly reflects the needs of employers – productivity, profitability and competitiveness – to the exclusion of social reproduction concerns at the levels of workers and their families. Within this framework, employee voice has largely been restricted (Hyman and Summers 2004), such that 'flexible working seems to be more concerned with accommodating life to rather demanding and unquestioned working hours rather than one of reorganising work to allow time for domestic and caring responsibilities' (Perrons 2003: 69). A further irony is the willingness of employers to pay to alleviate their employees' stress but their reluctance to provide facilities that might prevent it in the first place (Hogarth et al. 2001). In short, 'unless we consider both business and social imperatives, optimal outcomes cannot be reached' (Lewis et al. 2003). Herein lies the central relevance of the mutual gains perspective developed here, in which the gendered work-life experiences and preferences of workers are analysed alongside firms' economic requirements from the outset (following Maxwell and McDougall 2004: 390; Rapoport et al. 2002; Wise 2003: 35). This sequencing is crucial to avoid producing an analysis that might otherwise obscure (or worse perpetuate) problems of work-life conflict. This, whilst also recognising the pragmatic need to frame calls for socially progressive corporate change in a language that managers (and shareholders) can hear (see Froud et al. 2000).

Employer-provided WLB arrangements are typically grouped across four categories that (i) provide greater flexibility in scheduling *when* work is done whilst not decreasing total work hours (e.g. flextime, annualised hours, compressed work weeks); (ii) provide greater flexibility in the spatial location of *where* work is done (e.g. working from home one or two days per week); (iii) reduce total work hours (e.g. 80% FTE contracts); and (iv) provide assistance with childcare. As identified in Chapter 5, there is no single WLB panacea: workers' preferences vary not only within gender groups (by job function, department, household situation) but also for individual workers over the lifecourse, underscoring the need for more comprehensive bundles (Perry-Smith and Blum 2000) of WLB provision by employers.[1] In addition, the highly personal and subjective experience of work-life conflict means that workers' lived experiences of improvements in the quality of their personal and home lives that result from using different work-life arrangements in practice are similarly varied. However, against this diverse backdrop, the interviews point to significant reductions in negative work-to-home spillovers, particularly amongst workers in IT firms with more comprehensive bundles of WLB provision.

First, participants highlighted a reduction in the intensity and frequency of 'time-based conflict', in which time spent by individuals doing paid work otherwise reduces time available for carrying out other family responsibilities and personal life activities. There are a number of key components to this dimension of reduced work-life conflict. As a function of increased spatial flexibility of work through homeworking, participants described the benefits of a reduced commute for enabling more time (and a greater predictability and regularity of that time) for fulfilling family responsibilities and engaging with children's activities (e.g. sharing family meals, helping children with homework, playtime, bedtime reading and family outings). As one male software developer with two young children pointed out, 'the hour that I'd be in the car now I can now spend with the kids'.[2] Likewise, a male colleague who had been working at home five days a week for three years remarked:

'I joke with people that working from home's great, you sometimes hit a little bit of commuting traffic on the stairs but generally the journey is pretty good. My boys are four and seven, it's given me a huge amount of flexibility to actually be with them. Working in the office I was leaving home at 6:00 in the morning and wouldn't be home until 7:30 in the evening. I would have spent half an hour with them on a daily basis whereas this way, I can have my lunch with them, have my breakfast with them, it's far better.' Senior Software Developer, male, two children, US-owned IT MNC, Dublin

Technology workers also highlighted the advantages gained through their use of flextime and homeworking in enabling personalised workday time structures, in which they are better able to fit discrete packets of work around everyday activities of household reproduction and care. Examples of this work/life alternation – and the quicker periods of task changeover enabled by flexplace – include one female Senior Marketing Manager in Cambridge who will 'put on a load of washing and start work at say 9:30, and then work through 'til say 5 o'clock, and then I'll be doing the children, and then I'll get back on email around 7:30 or 8, until 10 pm'. Similarly the CEO who 'will take a break if I'm working at home to have lunch with my daughter, and then I'll go back to work and stop working at about 5:30, to spend time with my kids again until they go to bed, and then I end up working another shift after they go to bed with the US office, so we end up having another sort of 8:00 to 9:30 time period' (female, two children, IT MNC, UK SE region). For other workers, the use of flexible work start and end times was identified as enabling a more efficient use of time, to schedule everyday caring activities at times when they can be completed most quickly and efficiently, outside of the typical early mornings, evenings and/or weekends model of time for care dictated by 'standard' 9–5 working arrangements:

'I can't imagine there is joy to do laundry or ironing or go shopping on a Saturday when all the amateurs who work full-time are out. If you're working part-time you can be in the shops at 9 o'clock on a Tuesday or Thursday, it just works so much better. I think it would just suck you dry to have to have both of you work, and then do the whole kids thing and focus all your attention on them at the weekend. So we have a really great quality of life.' Recruitment Manager, female, two pre-school-age children, 3-day work week, IT MNC Dublin

And for other workers without children, the use of WLB arrangements giving greater temporal and spatial flexibility around paid work was credited with perceived improvements in health outcomes and reduced stress as a function of making more time available for sports and exercise, more space for hobbies and after-hours learning, better sleep patterns and more time to maintain friendships.

A second major dimension of the improvements in quality of life that result from IT workers using different WLB arrangements relates to perceived reductions in 'strain-based conflict' (Greenhaus and Beutell 1985), in which home and family interactions are otherwise negatively affected by stress and anxiety resulting from activities in the paid work domain. The most common statements within this category centred on increased compatibility of family and work commitments through reduced feelings of work-induced guilt, reduction of previously experienced 'care deficits' and associated negative impacts on children's well-being, and reductions of previous constraints on workers' abilities to attend key child events (parties, sports days, others) due to the primacy of work commitments. These reductions in strain-based conflict were most commonly expressed by working mothers trying to balance the demands of a busy IT career with motherhood (typically through a reduction in the demands of paid work, often through reduced hours working):

'When I had my first baby, she was in the crèche full time and it broke my heart leaving her in. I felt guilty when I was at work and I felt guilty when I was at home with her. Also, it was a very long day for my daughter, and I'd be driving at break-neck speed across the city, literally in tears if I was caught in traffic because she'd be there until 5:45 instead of 5:30. So working three days a week, I'm not guilty when I'm at work and I'm not guilty when I'm at home. So, for me it's absolutely fantastic.' Recruitment Manager, female, two pre-school-age children, 3-day work week, IT MNC, Dublin

Closely related to the reductions in feelings of guilt as identified here, several participants also acknowledged the role of WLB arrangements giving greater temporal and spatial flexibility of work, alongside reduced total work hours, in enabling an increased quality of time spent with children and reduction in the need for, and use of, outsourced care ('commercialised intimacy'), whilst also maintaining professional work interactions. This was particularly apparent amongst dual-career couples who both work in the IT sector, including one senior manager in Cambridge whose 'husband just changed job and got to choose his own office so he chose one close to where he lived. Having him at home in the mornings, he's there with the kids, it's just amazing his relationship with the children, how much it's improved. He could go two weeks before, but wouldn't see them except for the weekends.' Similar improvements were also evident in Ireland, including another female manager for whom:

> 'It's just extraordinary being able to spend quality time with the kids rather than just being exhausted all the time. And you also get to talk to grown-ups two days a week, which is wonderful, to be respected and to do a job that I love. You get out, you get dressed in normal clothes (rather than a tracksuit and a stained t-shirt), and you present yourself in a different way than you are as a mum. So that's great and it works on all sides.' Marketing Manager, female, two pre-school children, 3-day work week, IT MNC, Dublin

Several participants also described the cost savings to their households of remaining at home one or two days a week, which, despite a pay cut, gave rise to financial savings overall when offset against high costs of private childcare and lengthy commutes, particularly in the Dublin context. However, it is important also to note that the positive outcomes as described above by some female IT professionals around reduced guilt and making more time to care do not necessarily challenge ongoing gender inequalities of care *per se*, merely women's abilities to juggle an assumed majority share of childcare with paid work. Likewise two female technologists for whom reduced hours working had not yielded any meaningful reduction in strain-based conflict – this as a consequence of their continuing to do five days' worth of work activity in four days, with the other half of this double whammy, a reduction in pay.

More encouraging, however, a third dimension of improvements in quality of life experienced by workers as a result of WLB provision by IT employers relates to reduced 'cross-over stress'. That is, the challenges that otherwise result when work-life conflict experienced by one spouse impacts on the other spouse, as the latter increases their own unpaid work in response to the paid work demands placed on the former (see Pittman et al. 1996). As part of this, some female IT workers identified their male partners' beneficial use of homeworking and flex-time as leading to perceived reductions in their own experiences of cross-over stress, as these WLB arrangements enable their partners to undertake a greater share of everyday household social reproductive tasks (e.g. school run, shopping, cleaning). In this way, WLB provision by IT employers can be seen in some cases as helping to foster more gender-equitable household divisions of care.

Enhancing Firms' Innovative Capacities through Work-Life Provision (Mutual Gains, part II)

'The "work-life balance business case" might better be seen in a mutual gains kind of perspective, rather than simply in terms of business bottom-line. There's a perception on the part of employers or providers that these things are simply a cost. And if we can argue cogently that that's not necessarily the case, it clearly opens up the ground for us.' Director of Research, Irish Equality Authority, Dublin

Despite the multiple benefits that accrue to workers and their families as identified in the previous section, many employers remain sceptical of the benefits of this 'additional administrative burden'. At the same time, analyses of the organisational benefits of WLB provision to date have restricted their focus to output measures of firm performance: especially output per employee, output per hour, absenteeism and/or operating costs. While these are centrally important measures of 'revealed competitiveness' (Gardiner et al. 2004), they say little about the underlying sources and determinants of firms' abilities to compete through technological learning. As identified in Chapter 2, these determinants include firms' abilities to access external sources of information and knowledge, and then to reconfigure, transform and use that new knowledge to commercial ends (firms' 'absorptive capacities') in the development of new technological capabilities, new products and/or new ways of organising production processes and service delivery. These learning and innovation processes are not discrete, abstract, disembodied phenomena. Rather, they are *peopled*, by workers with gender identities, family commitments and extra-curricular interests, who interact on an everyday basis to combine varied skills, competencies, ideas and prior experience to create new knowledge and apply it incrementally in the pursuit of improved economic performance. As such, these processes are unavoidably shaped by labour geographies and the workplace institutional environments of which WLB arrangements form an increasingly important component. Nevertheless, this is an area in which conventional WLB business case analyses remain silent. Likewise an ongoing myopia within the regional learning literature means that economic geographers have largely failed to locate activities of knowledge production within the wider networks of social reproduction and care which unavoidably shape them in powerful ways.

In response, this section explores the learning and innovation advantages that can also accrue to knowledge-intensive firms through the provision of WLB arrangements identified by their workers as offering them meaningful reductions in work-life conflict. Table 6.1 shows that the most widely available types of employer-provided work-life arrangements amongst IT firms in Dublin and Cambridge are flextime (67%), working from home one or two days a week (66%) and part-time working (54%). Table 6.1 also shows worker take-up of different employer-provided WLB arrangements across the IT firm sample.[3] As an

Table 6.1 Employer provision and worker take-up of (formal) work–life arrangements, Dublin and Cambridge IT firm sample (2008).

Formal WLB arrangement	Dublin N = 74		Cambridge N = 76		Combined N = 150 (8068 workers)	
	% firms available	Ave % workforce take-up	% firms available	Ave % workforce take-up	% firms available	Ave % workforce take-up
Flextime (flexible start/end time, sometimes with core hours)	73	50	61	77	67	62
Flexplace (work from home 1 or 2 days a week)	74	18	58	49	66	32
Flexplace (work from home 3 or 4 days a week)	35	33	53	55	44	46
Job sharing (one job undertaken by 2 or more persons)	9	2	4	52	7	17
Annualised hours	8	23	11	50	9	38
Part-time work	49	15	59	33	54	25
Compressed work weeks (e.g. 5 days' work in 4)	31	27	29	48	37	30
Term-time working	8	14	9	20	9	17
Extra-statutory maternity leave*	32	33	12	42	22	35
Extra-statutory paternity leave**	14	8	9	28	11	16
Career break/sabbatical	19	14	7	32	13	19
Employer-subsidised childcare	4	7	8	3	6	4
Information referral service for childcare	4	0	3	0	3	0
Workplace nursery	3	0	1	0	2	0
WLB counselling/training	15	14	4	67	9	25

* % take-up calculated using companies' female workforces only

Table 6.2 Consistency of manager-perceived WLB learning benefits with measured improvements in firm performance (2004–2007).

Employer-identified benefit of WLB provision (2004–2007) N = 142	Ave measured change in firm performance (2004–2007)			
	Worker productivity (£ revenue per head)	Revenue (£)	% female workforce	Labour turnover
Increased worker productivity (61% firms)	+15,995	+302,563	–	–
Improved retention of women post-maternity leave (52% firms)	–	–	+5.3	–2.0
Increased workforce diversity (44% firms)	–	–	+3.6	–
Better environment for learning and creativity (54% firms)	+44,635	+817,082	+4.0	–2.2

aggregate consequence of these various levels of WLB provision and take-up, 61% of managers responding to the IT employer survey indicated improved worker productivity as a result of their total bundles of WLB provision and take-up, and 54% 'an improved corporate environment for learning and creativity'. These perceptions are typically more common amongst managers in firms with more extensive bundles of total WLB provision and higher rates of WLB take-up. Importantly, they are also consistent with measured improvements in firm performance across a series of metrics over the same time period (2004–2007; see Table 6.2). Only three firms identified a negative impact of WLB provision on their 'corporate environments for learning and creativity'. However, these same firms concurrently also demonstrated an average £20,500 revenue increase and 10% increase in female workforce.

In order to understand the mechanisms which underpin these aggregate patterns at the firm level, the analysis below draws on insights from 68 in-depth interviews with software programmers, testers, analysts, architects, team leaders, managers and other industry commentators, triangulated with survey evidence from over 300 IT workers (2008, N = 162 and 2010, N = 147). Research participants were asked to identify and explain the ways in which their take-up of different types of employer-provided work-life arrangements (as well as by their colleagues) impacted on their everyday abilities to function effectively as IT professionals. Interviews focused specifically on changes to what, where, when, how and with whom everyday work is done. Based on these insights, five mechanisms were evident in relation to WLB and learning outcomes within firms, and these are summarised in Table 6.3. Much more than a narrow focus on the implications of work-life provision for firms' strategic learning through R&D, the analysis explores their implications for the 'routine' or 'everyday' work practices, team

Table 6.3 Everyday mechanisms of work-life/learning advantage.

Mechanism	Worker-Identified Components	Interventions to Reduce Constraint
	WORK-LIFE LEARNING ADVANTAGE	
1. WLB and increased temporal and spatial autonomy of work	• Greater choice about where and when to schedule and accomplish work tasks. • Longer windows of uninterrupted time for better concentration and creative problem solving. • Self-perceived improvement in worker concentration, focus, motivation and engagement. • Retention of workers with significant, e.g. caring, commitments who would otherwise leave.	
2. WLB and increased work team heterogeneity	• Increased work team demographic diversity (gender, age, tenure). • Enhanced work team competencies, experience and perspectives for problem solving. • Work teams can draw on more diverse cross-firm networks of personal contacts – improved basis for benchmarking ideas. • Development of products in tune with demands of diverse customer base.	
3. WLB and enhanced sustainability of learning	• Reduced work overload enables more effective thought process with reduced error component. • Reduced stress, fatigue and burnout as fewer constraints on workers' abilities to think laterally and creatively. • Reduced stress enabling better lines of communication with colleagues.	
	WORK-LIFE LEARNING DISADVANTAGE	
4. WLB backlash, distrust and constrained worker agency	• Distrust/resentment among team members undermines group problem-solving ability. • Lesser willingness of workers on non-standard contracts to risk mistakes in pursuit of learning (stronger sense of scrutiny and fear of penalty). • Lesser willingness to challenge opinions of non-WLB-supportive supervisors/team leaders.	• WLB training seminars. • Peer mentoring and celebration of role models beyond masculinist, unattached 'ideal worker'.
5. WLB, fragmented work teams and limits on group communication	• Reduced spontaneity and informality of interaction. • Reduced frequency of work group communication: 'out of sight, out of mind'. • Difficulties scheduling team status meetings. • Difficulties scheduling customer meetings (developer–user learning constraints).	• Instant messaging, video conferencing, shared email, web-based project repositories. • Maintain work group capacities for multi-directional, simultaneous cycles of discussion, interruption, repair, feedback and learning quickly at a distance.

interactions and identities through which workers combine varied skills, competencies, ideas and prior experience to create new knowledge and apply it incrementally in the pursuit of learning, innovation and improved economic performance.

(i) WLB and increased self-determination of work

The core tasks of software development include systems analysis, software design, programming, testing, negotiations with users, and communication and cooperation within software teams. Accordingly, and in common with other knowledge-intensive sectors, IT work is characterised by the need for collaboration and interaction, combined with individual cognitive activities such as analytical thinking, conceptualisation and coding, which require significant periods of non-interruption during which to concentrate (Perlow 1999). Within this context, research participants emphasised the learning advantage of WLB arrangements which give IT workers increased autonomy, freedom and independence over the spatial location (e.g. teleworking) and temporal pattern/ sequencing of their work hours (e.g. flextime, annualised hours), in a manner that better enables them to engage in individual cognitive activities and to use their time more effectively in less fragmented blocks. The significance of this time sovereignty was contrasted with the negative outcomes of being expected always to work in a particular worksite location alongside other colleagues, and why that is often inefficient for individual creative work due to interruption and unnecessary repetition of analytical tasks. More than simply limited to smaller offices where some technologists 'struggle to think hard and shut out noise, amongst a shed load of distractions, to focus' (CEO, male, IT SME, Cambridge), this also exists in larger firms:

'It's when you take a piece of paper and it's got all these squiggles and designs on it and then translate that into a piece of code, from a concept to a physical item, that takes a lot of effort mentally. So if someone interrupts you halfway through an algorithm and you haven't put the basic outlines of it in, then you're going to end up in a right mess. You come back to it two or three minutes later having picked up a phone call and have to go all the way back to the beginning of that train of thought to be able to go over that process. Only for someone else to probably cut in again at another point two minutes down the line!' IT Engineer, female, US-headquartered MNC, UK SE region

In response to these difficulties, participants outlined the benefits of allowing them to do creative work when they perceive themselves to work most effectively, at times which often do not coincide with normal office hours rooted in an earlier model of industrial time. In multiple cases evenings were identified as the key period when 'I do my best thinking, either because it's the end of the day, therefore I have a deadline, or because it's silent in the office and no-one interrupts me' (CEO, male, IT SME, Cambridge); 'I might have been trying to do something all day, and then I'll spend a couple of hours in the evening and it'll come together perfectly. So it suits me to work later and then come in the next day at eleven or twelve' (Software Designer, female, IT SME, Dublin). And at the other extreme of the working day, 'I might have a piece of work that would take me three hours to do during the day. But in the mornings really early, by 5 I'm on fire, literally, and I can do it in 20 minutes. I'm just really productive early in the mornings' (IT Consultant, female, IT SME, UK SE region). Indeed, these time-sovereignty effects were also identified by some IT managers at the level of development work teams which 'had *total* flexibility of when they worked, all they had to do was deliver. And they would get ideas at 4 o'clock in the morning and they would come in. Sometimes they'd work for 12, 14, 16 hours at a stretch and then you wouldn't see them for three days, but that was just the way they worked and that was how we got the best out of them. I'd never seen that before, but it worked, it definitely worked' (Director of Human Resources, male, two children (middle school), IT MNC, Dublin).

And much more than IT workers just 'being precious', participants also pointed to the ways in which time sovereignty offers advantages for interaction in geographically dispersed international work teams. So while some often 'end up staying late, which works really well because a lot of the team are in New York and not here' (Software Engineer, female, MNC, UK SE region), others head in early 'to talk with Asia, like six or seven [a.m.] and when it comes to 12 o'clock I am exhausted, but I'll have a much more productive day than doing a nine to five' (Software Systems Engineer, female, MNC, UK SE region).

Preferences for increased *spatial* flexibility of work were also often justified on the basis that 'I can't work at work!' (see Gregg 2011). And whilst certainly not immune from technological challenges of 'what is possible' becoming 'what is expected' (e.g. in terms of logging in after hours), research participants engaged in regular homeworking in pursuit of an improved WLB nevertheless outlined the benefits for the quality of their everyday work performance in terms: of fewer distractions and longer windows of uninterrupted time for immersion, focused concentration and creative problem solving. The ways in which employer assistance with childcare and flextime helped some workers to reduce their previous need for, and distractions of, juggling childcare informally during work hours were also highlighted in this context. However, for several homeworking female IT professionals with children, the learning gains they identified as an outcome of working from home were crucially also dependent on childcare assistance through the use of nannies and wider networks of family support, which thereby enabled them to 'import industrial time into the home' in a manner previously identified amongst men (Sullivan and Lewis 2001: 141):

'A couple of times I've had a big chunk of stuff to write, I'm in an office with two people so that can be easier to do at home. But I have to tell the nanny to take the kids out, you know, because even upstairs you can hear all the mess. So certain days I would have dropped them at crèche, come home, done a load of work, gone off and picked them up and you get a lot of work done there.' Marketing Manager, female, two pre-school children, 3-day work week, US IT MNC, Dublin

As such, these everyday work-care arrangements underscore further how geographers' widely researched 'regional worlds of knowledge production' are crucially dependent on under-researched 'regional worlds of social *re*production'.

In sum, by making use of differently preferred work-life arrangements – effectively reshaping the temporal and spatial boundaries between home and work – IT workers are able to effect changes in their workplace learning environment, which results in a self-perceived improvement in their concentration, focus, motivation and engagement. This was variously described on the basis that 'I don't hate coming into work anymore' (Software Engineer, female, US-owned IT MNC, UK SE region), and 'just the kind of general feeling of just feel more positively towards the company, if they're willing to go out of their way for you, you do go out of your way for them' (Technology Lead, female, IT MNC, Cambridge). Similarly:

'I run in here on a Monday morning, absolutely love it, thrilled to be back here and charge out again on a Wednesday and say "that's great, see you again next Monday!" The joke is that you get to see steam coming off the top of my head, you know, and everybody heaves a sigh of relief when I'm gone on Thursday because they feel they can work at their own pace.' Software Business Development Manager, female, two young children, 3-day work week, US-owned MNC, Dublin

These insights usefully extend earlier work in economic geography which suggested that to learn and innovate effectively, workers must feel motivated and engaged (Benner 2003). In conceptualising these impacts it is also useful to recognise parallels between these workers' personal testimonies and established research literatures beyond economic geography – in organisational and social psychology – that demonstrate how learning and the generation of ideas are enhanced when motivated individuals and teams have relatively high autonomy in the day-to-day conduct of their work (see, e.g., Bailyn 1985; Amabile and Conti 1997; Osterlow and Frey 2000). In so doing, this work moves beyond a narrow focus on the *personal* characteristics of creative individuals to explore how different *organisational* structures (including work-life provision) can also enable and constrain the ability of individuals and groups to work together in the development and application of new ideas.

(ii) WLB reducing gendered exclusions, enabling heterogeneous work teams

The second dimension of work-life learning advantage concerns the consequences of employer-provided work-life arrangements for increasing firms' repertoires of perspectives, skills and networks of connection (or what Lundvall and Johnson (1994) have referred to as 'know-what', 'know-why', 'know-how' and 'know-who'). Notwithstanding the regular need for individual cognitive activities as outlined above, firms' larger innovative capacities also benefit from interaction among workers with a range of technical expertise, knowledge and experience. When individuals with diverse and partially overlapping competencies come together and collectively seek to articulate their ideas about a new product or technology they are trying to develop, they are forced to clarify those ideas and to derive more adequate concepts and models (Lawson and Lorenz 1999: 312). This in turn allows ambiguities in the perceptions of the individual partners to surface, and provides a basis for comparison (Oinas and Malecki 2002). Or as Rosabeth Moss Kanter usefully has put it, 'contact with those who see the world differently is a logical prerequisite to seeing it differently ourselves' (1988: 175). The natural conflict that emerges from the interaction of workers' different perspectives, accumulated experience and insights ensures that a wider range of possible solutions is considered, with an increased potential for the development of new and unexpected ideas (e.g. Pelled et al. 1999; Reagans and Zuckerman 2001). As such, work team diversity has been widely identified as a necessary precondition for successful innovation and corporate capacities for comprehensive problem solving in the face of new events.

Significantly, the interviews highlighted how comprehensive bundles of work-life provision – and employee uptake of those WLB arrangements – can enhance work team diversity. This by means of retaining experienced IT professionals with significant caring responsibilities and commitments outside work, some of whom had previously considered leaving particular firms – and in some cases the IT sector altogether – due to recurrent problems of time-based and strain-based work-life conflict. In support of this mechanism, over half of all managers responding to the employer survey (N = 150) indicated improved female worker retention post-maternity leave as a result of their total bundles of WLB provision and take-up, and 44% increased workforce diversity. This includes the extreme case of one large IT multinational in Dublin whose managers expressed alarm at discovering over 50% of their employees returning from maternity and/or parental leave were no longer employed at the company just one year on, a worrying trend which they attributed to a previous lack of adequate work-life support.

Reinforcing these claims, several human resources managers also emphasised this skills retention component of the WLB business case over other moral arguments for work-life provision as outlined in Chapter 3:

'Yes, we feel it the right thing to do, but that actually isn't the only reason we're doing it [WLB provision]. It's not a charity, things like advancement of women, we feel we need to do that because we need to exploit that part of our workforce, in a world where skills are scarce, trying to make sure you're exploiting those. So there's a business driver with work-life balance.' Human Resources Manager, female, US-owned IT MNC, Dublin

Indeed, one female participant also pointed to a more direct link between motherhood and its skills advantages for IT employers, on the basis that 'although I am in Toddlerville three days a week, they're incredibly good negotiators, so to go then into that kind of customer-facing role where they want everything and they want it right now, it's exactly what you've been dealing with for the last five years at home!' (Sales Manager, previously Software Engineer, IT MNC, Cambridge). Several participants also pointed to the learning advantages that accrue to IT firms as a function of work team gender diversity enabling the development of 'software products more in tune with the demands of a user base that is itself 50% female!'

So, in contrast to popular conceptions of WLB provision as merely an additional administrative burden to the firm, work-life provision was identified as enabling firms to accommodate and retain a workforce with a diversity of extra-curricular responsibilities and/or caring commitments, genders, ages, positions in the lifecourse and life experiences. And while one participant was adamant that 'young all-male teams are simply more effective' (!), a more general consensus emerged around the everyday learning benefits of working in *less* homogeneous work teams with a greater diversity of opinion and accumulated experience to draw on:

'If you've got five people maybe around the same age, the same culture, they'll probably come out with some fairly similar ideas. But if you've got a lot of people with different ideas, you can manage those different ideas for innovation, the creativity. So let's bring people in because they're different and use that experience ... I mean it's a business motivation, you know, I'm not here [in HR] because it's nice to have diversity, I'm here because it affects the bottom line.' Diversity Manager, female, two children, IT MNC, Dublin

These everyday insights from IT managers and workers are consistent with studies of work environments in other sectors characterised by high levels of uncertainty, in which demographic diversity is positively correlated with superior corporate performance (e.g. Eisenberger et al. 1990; Díaz-García et al. 2013). Underpinning this learning advantage is the recognition that attitudes, cognitive functioning and beliefs are not randomly distributed in the population but tend to vary systematically along a range of demographic variables including gender, ethnicity and age (Robinson and Dechant 1997). (Or as one female CEO in my study summed it up, there is a 'huge value in employing individuals who have been round the block'.) Additionally, the quality of reasoning in majority opinions is enhanced by the existence of consistent counter-arguments from members of minorities (Nemeth 1986), such that work groups think in more realistic and complex ways (Milliken and Martins 1996). In this way, work team diversity reduces the chance of 'groupthink' – or what sociologists refer to as 'occupational psychosis' or 'trained incapacity' – which is the tendency for teams who interact only with others with a similar outlook and focused on one area to become less able over time to learn new things (Kanter 1988: 176). Further reinforcing these learning advantages, several research participants also pointed to how firms' work-life interventions to retain highly qualified IT professionals with significant caring commitments also enabled them to tap into larger networks of knowledge and expertise ('know-how' and 'know-who') outside the boundaries of their respective firms. These include the cross-firm networks of female IT professionals through which this study was in part conducted: Girl Geek Dinners, Women In Technology and WITS Ireland.

(iii) WLB, reduced stress and fatigue, and enhanced sustainability of learning

'There was no doubt that I was heading towards a burnout – which most people at some point hit when they're at [large US-owned IT MNC] because it's quite relentless. It's non-stop, you finish one thing, you run headlong straight into the next. I just was turning into a little bit of a zombie.' Principal IT Consultant, female, IT SME (previously at large US-owned IT MNC), UK SE region

Time-based competition has long been identified as a core defining feature of high-technology industries, in which firms seek to bring products to market faster than their competitors (see Stalk and Haut 1990). For workers, a certain degree of time pressure and manageable stress has also been identified as spurring innovation through a general sense of urgency to accomplish important work tasks (Amabile 1996; West and Sacramento 2007). However, the results from this study also point to the longer-term disadvantages of overwork and long hours work cultures for constraining innovative capacity. The third mechanism of work-life learning advantage therefore relates to the role of employer-provided WLB arrangements in fostering an increased longer-term sustainability of worker learning.

A range of phrases and labels were used by research participants to describe the intense patterns of working in the IT sector – and associated feelings of stress and fatigue – ranging from 'burnout', 'death march' and 'crunch mode' to the more colourful 'working our nuts off'. Research participants described the ways in which time pressure and long hours working can also result in self-identified reductions in their abilities to work effectively and to engage in the everyday activities of problem solving, coordination and symbolic analysis which define their learning work as IT professionals. As described by one female IT engineer in Cambridge, 'when they put deadlines down that are ridiculous, and you pretty much do a 24-hour day, just to get a product out the door, you don't think as laterally, you don't think as logically'. Similar learning constraints were also identified in Dublin, where technologists explained that 'in IT you are literally buying the brain power of your consultants and engineers' and that 'as stress levels go through the roof, their creativity, their problem solving is just going right down'. In short, by repeatedly imposing too great a workload in too short a timeframe, the motivational benefits of time pressure in the short term often quickly diminish.

Significantly, these admissions by IT workers are consistent with previous studies which have highlighted the dangers of high time pressure, overwork, stress and fatigue as constraints on creativity (e.g. Amabile and Conti 1997; Amabile et al. 2002; Maule and Edland 1997; Huhtala and Parzefall 2007) – even in some instances with effects similar to alcohol intoxication as a function of diminished memory, slowed reaction time, lapses in attention to detail and decreased concentration (Josten et al. 2003). Admissions by several participants that they deliberately cut back on sleep to cram in more work are also cause for concern, given other recent research that has connected fatigue and voluntary sleep loss to workers' reduced abilities to process information, deal with distractions and remember in the short term (see Chee 2015). Encouragingly, IT managers in Dublin and Cambridge were often critical of other 'tech places, who will turn every project into another big death march: long hours until it's done', instead suggesting that 'you can only pull that trigger so many times' (CTO, male, Irish IT SME, Dublin). Instead they made clear that 'it's actually in the company's interest: in harsh economic terms, how do you extract optimal value from that person? It probably isn't hammering them into the ground 50 hours a week until they collapse in a heap in their 40s!' (Principal IT Consultant, female, IT MNC, UK SE region).

Extending these observations, managers also highlighted the negative learning effects of prolonged time pressure, work-life conflict and stress for talented workers 'to begin to hate the place and leave' in 'companies where they're whipping people until they're so exhausted that they can't function the next week, or the next day' (CEO, female, IT MNC, UK SE region). The commonly identified result is a significant loss of accumulated skills, knowledge and expertise. In response, managers emphasised the importance of work-life balance provision as a means for enabling the longer-term sustainability of learning and innovation in their development teams. The aim, then, is to 'give everybody the time to regenerate, by doing your own social things, spending time with your family or just having time to sleep, that actually we bring back into the company – we're enthusiastic to be there and to get on with things. In the long run it makes us much more productive' (Head of Technology, female, US HQ MNC, UK SE region).

Several research participants also highlighted the role of WLB provision in helping to maintain improved lines of communication with other team members through a reduction of work-life conflicts and stress which otherwise undermine their perceived abilities to interact and communicate effectively with colleagues – this on the basis that 'when you're stressed out of your head, you can't come down, you can't communicate'.[4]

The results from the IT worker survey (N = 162) are nevertheless striking, with an impressive 94% of workers indicating that, as a result of making use of their personally preferred employer-provided WLB arrangements, they were less stressed at work. In addition, 79% reported greater engagement with their work; 78% that they could think more creatively at work; and 75% that they could learn more effectively at work. Importantly, these figures are even higher for working parents with young families (n = 48), with equivalent figures of 82% for greater engagement, 84% thinking more creatively and 78% learning more effectively at work. The larger point is that in combination with the two other mechanisms of work-life learning advantage explored above, employer-provided WLB arrangements have powerful – yet under-researched – implications for firms' absorptive capacities in terms of their enabling firms to begin to challenge gendered constraints on firms' everyday capacities for learning and innovation.

WLB Provision Creating New Challenges for Everyday Learning and Innovation

In addition to the learning benefits that arise from employer-provided supports for improved work-life balance amongst their workforces, a number of challenges also arise. At the broadest level, these challenges were identified in relation to WLB arrangements which give workers greater *spatial* flexibility (specifically 'from you or your colleagues regularly working from home'), with 17% of workers identifying difficulties scheduling meetings with customers, 27% limits on effective communication with colleagues and 18% reduced group problem-solving abilities. The same three problems were also identified in relation to 'you or your colleagues regularly working reduced hours', with 20% of workers identifying limits on effective communication with colleagues, 16% difficulties scheduling meetings with customers and 12% reduced group problem-solving abilities. These challenges also have a significant gender component, on the basis that working mothers experience work-life conflict most acutely and are therefore more likely than other worker groups to seek alternative arrangements in the first place (Houston 2005; Gregory and Milner 2009). It is important, therefore, for economic geographers to explore these difficulties and technological solutions for their reduction.

(iv) WLB backlash, distrust and constrained worker agency

Firms' innovative capacities are not randomly distributed but are dependent on a distinctive set of corporate cultural traits that promote the successful development of new ideas, new products or services, new technological capabilities and/or new ways or organising production. These include a climate of openness in which debate and constructive conflict are encouraged; a willingness to break with convention; widespread support for trying new things; and the right of employees to challenge the status quo (Deal and Kennedy 2000; DiBella et al. 1996). Firms' abilities to innovate thus presume a necessary relationship between learning and active employee involvement at all levels; that all employees can act as independent learning agents, experiment, respond creatively to new conditions and make mistakes as they learn without penalty (Spender 1996). Work environments that promote and accept informed risk-taking enable learning because the innovation process is uncertain, such that hoped-for timetables may prove unrealistic and not match the actual pace of progress (Kanter 1988: 171). Most notably, AnnaLee Saxenian's (1994) work emphasised workers' willingness to risk failure as a key component of Silicon Valley's capacity for learning and innovation, divergent from risk-averse behaviours and the stigmatisation of failure in Boston's Route 128.

Against this backdrop, however, the results here point to a *differential* ability of workers to act as independent learning agents and to have the confidence to take informed risks and make mistakes. These constraints stem from some workers' feelings of vulnerability associated with their take-up of non-traditional working arrangements in pursuit of an improved WLB, of which they perceived managers and co-workers to be less than supportive. Participants variously described these perceptions as ranging from feeling 'under-valued', to managers 'doing me a favour', to 'feeling like a lesser citizen'. In some cases, participants also described more extreme feelings of resentment from co-workers:

'I was talking to a woman recently, she's the only single person without a child in her department, and she feels she gets nailed every single time. She was quite resentful towards it, and felt "Look, I'm paid the same amount as them, I don't get any praise or benefit for it, and it's not my fault that they have children", she was extraordinarily resentful towards it, I was quite taken aback.' Human Resources Manager, female, US-owned IT MNC, Dublin

The point is that employees behave in certain ways based on 'the knowledge that someone who knows and cares is paying close attention to what we do … and is likely to care when things aren't going according to plan' (O'Reilly 1989: 11). And this mechanism is particularly significant in the IT industry as a function of decentralised management structures in which autonomous work teams are disciplined by subtle yet strong systems of normative control through peer pressure (see Kunda 2006). Here, the interviews pointed to a clear perception amongst several workers engaged in homeworking, flextime and reduced hours that their behaviour is being *more* closely monitored than colleagues whose commitment to the company is regarded as unproblematic as a function of their constant physical presence in the office during 'normal' work hours. Contrast this with multiple instances of managerial and co-worker scepticism around WLB provision and worker take-up in an industrial context where the 'ideal worker' has long been defined in terms of presence-as-commitment:

> 'The difficulty of managing people in a home-based or a non-office-centric environment [is that] it's not easy for me to assign someone and say, "Okay, you need to make 30 calls and by doing that I'll have evidence that you have been productive for the day". Clearly, this is all about trust. However, you and I know that trust comes with time, trust is a very fickle thing and can come and go with ease, and maybe I'm not the most trusting individual in the world.' CEO and founder, female with two children, IT MNC, UK SE region

> 'Maybe for junior staff it's a tendency for people to go home and think, "This is great, I can work at home and no one's watching me, I can do whatever I want, I'll go and ride my bike for an hour"'. Senior Software Developer, male, two children (school-age), home works full-time, US-owned IT MNC, Dublin

Strikingly, these comments contrast with the kinds of supervisory and co-worker relationships characterised by feelings of being 'valued', 'supported', 'trusted', 'encouraged', 'relaxed' and 'secure' identified in previous studies as more likely to enable creative achievement (Oldham and Cummings 1996; Janssen et al. 2004). In addition, some workers availing of alternative WLB arrangements in pursuit of reduced work-life conflict expressed an apprehensiveness about engaging in confrontation and challenging the opinions of supervisors and team leaders for fear of drawing further attention to themselves. Here, then, anxiety of threatening evaluation undermines the feelings of being valued and respected by colleagues, self-esteem and mutual trust that have long been identified as positively underpinning creativity and innovative capacity (e.g. Redmond et al. 1993; Vosberg 1998). Likewise the sense amongst some homeworkers that their competencies and skill-sets are not being fully utilised by managers in a manner that would otherwise offer the firm the greatest learning advantage, on the basis of being 'out of sight, out of mind'.

The results suggest, therefore, that female (and some male) workers making use of alternative/non-standard working arrangements experience lower levels of 'psychological safety' in terms of feeling 'able to show and employ one's self without fear of negative consequences to self-image, status or career' (Kahn 1990: 708).[5] In this way, firms' capacities for learning and innovation are unavoidably contingent on the relationships between workers, co-workers and supervisors (Mumford 2000). Indeed, the gendering of these dynamics also plays out in complex ways, as illustrated by one female HR manager at a large

US-headquartered IT multinational's Dublin office who made it clear that 'You just have to be careful not to have a department with a lot of women in it'. And further reinforcing that contingent learning relationship is the recognition of *customers'* resentment of 'abnormal' contracts, as identified here around part-time workers:

> 'We are, amusingly, off to court in 11 days to take a client to court who's refused to pay, claiming that the work was lacking, which is far from the truth. And in one of his emails he threw something along the lines of "You've done terrible work and the person managing my private work was working part-time". And I just thought it incredible that this individual felt that a part-time worker had severely compromised the project, and perceived that they were getting a lesser service and attention. But amusingly, this is an individual paying for our services for three hours a week and he's complaining that someone was only in for 20 hours a week!' CEO and founder, female with two children, IT MNC, UK SE region

Once again, the part-time worker whose work was in question here was female. In this way, WLB provision yields potential constraints on developer–user interactions, which have been widely identified in previous studies as underpinning firms' innovative capacities (see, e.g., Grabher et al. 2008). Through these interactions, developers with the technical knowledge base are able to gain a fuller appreciation of the demands and needs of users, as they move from passive recipients of technologies to active involvement in their configuration and evaluation. Managers, however, described their frustrations at having to deal with the challenges of WLB provision and uptake for managing such active developer–user relationships:

> 'It is a dynamic of whether in some cases you want to say to your customers that some of your employees are part-time. We tend not to raise it. You'd always like your employees to present themselves as full-time, that they're always available, rather than the customer saying, "Well [colleague's name]'s collecting her kids so maybe I won't ring her now at the moment", that type of thing.' CEO, father, four children, Irish indigenous IT SME, Dublin

> 'It's just there has to come a point where you say, "Are you working to move this business forward or is it a total convenience because it happens to fit very well with your family life?" People who have the commitments, such as pick up a kid, clearly will never stay a minute later. And I know it might sound terribly Orwellian or old-fashioned to be wanting people to stay later, but sometimes the unexpected happens, and stuff needs resolving fast, and the convenience of people going back to their other activities, their life, their loves, is not always convenient for the immediate goals of the business.' CEO, male, locally founded IT SME, Cambridge

The point is that the constraints identified in Chapter 5 in terms of colleagues' scepticism and managerial non-ratification in practice of formal WLB provision also give rise to a series of gendered constraints on firms' capacities for learning and innovation.

(v) WLB, fragmented work teams and limits on group communication

Communication has been variously described as the 'nervous system' or 'cement of innovation activities' (Kratzer et al. 2004). Face-to-face (F2F) interaction between co-workers is vital because it enables communication to occur on many levels simultaneously – verbal, physical, contextual, intentional and non-intentional (Storper and Venables 2002: 354–355). Non-verbal cues such as facial expressions, gestures, gaze and posture can enhance communication content by expressing relevancy or importance, urgency, validity or extent of agreement (Andres 2002). Face-to-face interaction therefore allows team members quickly to develop knowledge and understanding of each other, to adjust their individual and collective behaviours accordingly, and to reduce ambiguity and uncertainty arising from complex interrelated tasks (He et al. 2007). Within this context, research participants explored the implications of different types of WLB arrangements (particularly homeworking, reduced work weeks and flextime) for everyday interactions and patterns of communication between team members:

> 'In software product development, communication is what makes it work between the team developing stuff. There's the illusion of when everything's minuted and statused and the communication channels are often quite formal, but actually what makes things tick is often the casual and ad hoc communication. And if you're home or away you're out of that, you're not picking up that stuff going on. It's amazing how quickly you can have quite strong effects from very small changes in location.'
> CTO, male, indigenous IT SME, Dublin

At the broadest level, interviews pointed to feelings of marginalisation amongst homeworkers from face-to-face networks of interaction amongst their work teams, and negative impacts for the 'way that we work together, the way that we have a rapport, the things that we share, the ideas, we can be creative … I think we would lose that if we allowed everyone to start working from home, we'd actually be much poorer as a company' (Head of Technology, female, MNC, UK SE region). Workers suggested that 'If you do it longer term you get sort of forgotten about, you're not picking up that team learning dynamic', and that 'you don't have that "water cooler effect", you don't have that "Let's go grab a coffee. What are you up to? What project are you on?" to connect and bond'.

More specifically, workers identified the learning challenges of teleworking and reduced work weeks for the kinds of 'agile models of software development' adopted by multiple IT companies within their work teams. This family of software engineering processes rejects bureaucratic, micro-managed, slow models of development in favour of a 'lightweight' approach that reduces market risks associated with drawn-out product development life cycles. Agile methods are based instead around short iterations of one to four weeks, in which new functions are planned, analysed, designed, coded, tested and documented. The goal of agile development is to have a workable release available at the end of each iteration, when the development team also re-evaluates its project priorities (Highsmith 2002). Accordingly, agile software development methods emphasise a preference for regular peer face-to-face discussion over written documentation and status review meetings; close daily cooperation between analysts, systems engineers, architects, programmers, testers, managers and customers; and regular informal adaptation to changing circumstances and uncertainty ('you know, decide on what you're gonna do, and make changes on the fly' (CTO Dublin)):

> 'If you're working in the middle of a project, responding to changing requirements and it's a multi-disciplined team (business analyst, technical software developer and say a QA person), one of the parts of that is a daily meeting of the people to communicate … part of the agile development is that those people are actually physically sitting together and communicating very informally and very rapidly amongst each other to facilitate the short-term deliverables. The underlying proposition is that they're much more productive when they do that.' CEO, father, four children, Irish indigenous IT SME, Dublin

One problematic effect of teleworking, therefore, is its undermining the frequency (time spent communicating), formalisation (degree of prior planning versus spontaneity) and structure (direct or mediated) of information exchanges which support agile development.[6] As one CEO in Cambridge explained, 'the simplicity of speaking to someone two metres away from you and saying "I've got a problem here, can you come and have a look?" is a valuable communication process and doesn't always lend itself to remote desktop, telephone, chats and so forth. It's more about the requirement to have hour by hour interaction. Do you write someone an email? Or do you just ask the bloody question, get an answer and turn it into an interactive dialogue, and have the thing nailed within a couple of minutes rather than turning it into something that spans over several days?!' (CEO, male, locally founded IT SME, Cambridge).

Extending these work-life learning challenges, interviews also pointed to the challenges for communication, coordination, negotiation and work group functioning that emerge from reduced work weeks, identified by homeworkers and by their managers alike, and varying with the numbers of days per week spent away from the office:

> 'Three days a week crosses some sort of threshold where it begins to cause a lot of logistical problems: the permutations of what those two days where they're not working, people can never get their head around, or they themselves might swap this day for that day. And basically, after a while, no one knows when you're in and out. So it's less about the quantity of work and much more work team logistics, for synchronising and contacting and meeting face to face. It has a general wear and tear, a negative effect on the person and on everyone else's interaction with them.

Like a Friday and Monday, it means that somebody on Thursday doesn't see you 'til Tuesday and there could be status meetings and all this, that and the other going on.' CTO, male, indigenous IT SME, Dublin

And again, the result is that firms are unable to capitalise fully on the embodied skills and accumulated experience of some team members who are homeworking and/or working non-traditional hours.

Technologies for change? Reducing the learning constraints of work-life provision

'As an employer I find it hugely challenging and very difficult to manage a workforce that's not all in the office at the same time. Whilst it would be lovely to support everyone doing exactly what they want to do when they want to do it, some of my team at the moment need to redefine the boundaries of where the balance is.' CEO, male, IT SME, Cambridge

Importantly, the challenges to learning identified here as a function of workers making use of employer-provided WLB arrangements appear to be far from monolithic and inevitable in all firms. In relation to the first mechanism of constraint rooted in WLB backlash and distrust, some firms have implemented management training, showcased positive role models and developed peer mentoring to give employees appropriate work-life support, positive encouragement, help clarify role expectations and best practice, and recognise and demonstrate the value of working arrangements to the firm beyond the masculinist ideal worker model. Much more common, however, were interventions in relation to the second mechanism of constraint, centred on the development of information communication technologies which allow for face-to-face interactions *virtually*. The advantage of face-to-face interaction is that it offers the usual capacity for interruption, repair, feedback and learning, making it possible for two people to send and deliver messages virtually instantaneously (Nohria and Eccles 1992: 292). In order to mimic these advantages for homeworkers, some Dublin IT employers have experimented with online team meetings and brainstorming sessions using instant messaging and video conferencing facilities to maintain work group capacities for multi-directional, simultaneous cycles of discussion, interruption, repair, feedback and learning quickly *at a distance*.

Other relevant technologies include secure web-based project repositories where team members post task status information, data and documentation, and which are accessible to all team members whether at work or home. An advantage of these centralised web-based systems is that they overcome problems of team members leaving crucial files on memory sticks at work when working from home (or vice versa), or else saving files on a home or work computer only. Other research participants described the benefits of group email systems that help ensure the quality and timeliness of client interactions when team members work from home or work a reduced work week. For example, time-sensitive emails might be accompanied by text alerts, or else prompts linked to instant messaging software. And once a team member has replied to the client query (typically from a generic team email to provide continuity to the client), the email system automatically copies in all other team members, whilst simultaneously preventing anyone else late to the conversation replying to the same email again, and logs the communication string centrally for future reference.

'One of [our] major software brands is … all about software development: case tools, modelling, development tools, rapid development, source code management, all that type of thing. So the technology is available that would allow people to work remotely, check out a piece of code, do their work, check it back in, compile it and so on, in a remote type of fashion … to allow developers to work from home. I am sure that as that technology becomes more prevalent and adopted internally, we will start to see more people working from home.' Senior Software Developer, male, two children (middle school age), home works full-time, IT MNC, Dublin

Research participants therefore acknowledged the role of communication technologies in mediating the negative impacts of WLB provision on interactive learning and innovation processes, and hence for reducing tendencies to marginalise women and men with domestic responsibilities working from home. Significantly, the physical form of these technologies (and their visibility to children) also matters. As described by two different female managers in Dublin both working a 3-day work week, 'if something has to be actioned on a Thursday, Friday, then I just log on from home and do it. I've two small children, and it's just as soon as I open the laptop one of them starts pushing my legs away from it, she doesn't want me near it' (IT recruiter, two children pre-school). 'If I had my laptop open in the kids' playroom the kids would just go ballistic just trying to push everything and all that, and it just became unmanageable. Whereas this [smartphone] is less visible to them, you're walking them down the street, you can just do a couple of emails and replies, every couple of hours you check in, so it's actually much handier for me' (Marketing Manager, two children pre-school).

Of course, it is important not to position these technologies as some WLB panacea. Several research participants acknowledged the 'double-edged sword' nature of instant messaging, broadband and PDA devices in terms of their facilitating more effective integration and team interaction whilst working from home, but yet also making the already porous boundaries between 'home' and 'work' substantially more blurred. In some cases, workers described how they 'end up working harder than I really should be, because you're never really turned off, you can VPN in anytime, it is literally a click'. So then, 'you get emails in, just check this for five minutes, and three hours later you've got this whole set of coding that doesn't work and you're trying to debug, you always get sucked back in' (CEO, female, two children, IT MNC, UK SE region). Likewise for other workers, 'it means that my work and home life just merge into one – constantly doing a bit of both and still on my laptop long after the boys have gone to bed – so my social life suffers and I never turn off. But then I couldn't do a full-time job without this flexibility' (IT Consultant, UK SE region).

The ironic outcome of homeworking technologies, then, is that they risk contributing to *decreased* personal perceptions of an acceptable work-life balance. Extending these critical insights further, other workers commented on the danger of mobile communications technologies for WLB in terms of their enabling 'what is possible' to become 'what is expected' in order to appease customers, and/or to demonstrate commitment (see also Eikhof et al. 2007). Although, as one participant pointed out, 'it is possible to turn the phones and laptops off!'.

Work-Life Provision Rollback as Learning Disadvantage in the Aftermath of Recession?

The everyday challenges faced by workers struggling to juggle competing commitments of paid work, home and family remain stubbornly persistent and highly gendered. Yet despite the profound moral and social significance of work-life balance as a means for improving gender equity in market employment, facilitating a more equal gender division of household caring, and improving the well-being of workers, families and communities, employers remain unlikely to implement meaningful work-life arrangements unless they can identify bottom-line economic advantages that arise from their implementation. These problems are worsened by the challenges of growth in the wake of recession and ongoing – and indeed increased – employer scepticism of the business case for WLB. In response, this chapter has explored the learning and innovation advantages that result from work-life provision in knowledge-intensive firms, as part of a mutual gains WLB research agenda.

In terms of repositioning the WLB business case, the results suggest that by making available the kinds of WLB arrangements (particularly reduced work weeks and working from home) identified by workers as offering meaningful amelioration of everyday gendered work-life conflicts, employers can also enhance the kinds of learning and innovation processes that are widely recognised as fundamental to firms' long-term sustainable competitive advantage. These effects span three dimensions: (i) self-determination of the temporal pattern and spatial location of work yielding a self-identified improvement in workers' concentration, motivation, engagement and creativity; (ii) the potential to attract and retain a more demographically diverse workforce thereby widening firms' repertoires of perspectives and skills, networks of external contacts and hence capacities for comprehensive problem solving in the face of new events; and (iii) improved work-life balance, reduced fatigue and sustainable learning.

This is not to deny simultaneous *challenges* to interactive learning that result from team members working from home or working reduced work weeks. However, such negative outcomes are not inevitable. Rather, they are mediated by managerial buy-in and work group ratification and the use of information communication technologies to maintain virtual face-to-face interactions at a distance. Indeed, the combined effects of these mechanisms as identified by workers and managers were consistent with measured improvements in firm performance across a series of metrics over the same time period (including revenue growth, worker productivity, workforce diversity and labour turnover). And nor is this to deny work-life balance demands by some workers 'who want more work rather than less'. While these concerns were certainly expressed by several recent graduates, a core premise of this book is to increase workers' bargaining power for non-traditional working arrangements vis-à-vis employers who are sceptical of the bottom-line economic advantages that arise from their implementation. The case for persuading employers to allow some workers to work longer and harder for their own enjoyment seems somehow less urgent.

The analysis points, therefore, to the role of work-life provision as a significant – yet heavily under-researched – factor that can enhance the competitiveness of knowledge-intensive firms. In order to explore the extent to which this work-life advantage remains relevant in the context of an economic downturn, a second online survey of IT workers was rolled out November–December 2010 (N = 147).[7] Once again, the vast majority of workers surveyed identified the benefits of WLB provision for firms' learning and innovative capacities as: reduced stress (91%), greater levels of engagement with work (79%), being able to think more creatively at work (76%) and learning more effectively at work (73%). And yet, while the recessionary survey data identified several IT employers expanding their WLB provision in order to improve talent recruitment and retention, other firms were reducing their work-life provision in pursuit of short-term cost savings. Reinforcing these problems, 40% of survey participants also identified a reduced willingness *to use* their employers' formally provided WLB arrangements in practice, this compared with prior to the onset of recession. This was typically premised on fears of a lack of informal ratification by employers and negative career outcomes.

This subtle recessionary rollback of work-life provision is ironic given that it limits the very WLB arrangements which this chapter has identified as not only reducing incidences of time-based and strain-based work-life conflict amongst workers and their families, but also positively underpinning *firms'* learning and innovative capacities; which is even more vital in support of economic growth in the wake of an economic downturn. Also worrying, these patterns are consistent with stalled work-life provision and a shift in managerial attitudes regarding the locus of WLB responsibility from employ*er* to employ*ee*, as identified beyond IT in a wide range of other industrial sectors in the aftermath of recession (van Wanrooy et al. 2013; TUC 2012, 2015a).[8] Indeed, these shifting patterns of WLB provision also have implications for learning dynamics *between* firms. These form the focus of Chapter 7.

Notes

1   Previous studies of the organisational outcomes of WLB provision have advocated an analytical focus on total bundles of WLB provision on the basis that it is impossible to delineate the impact of one single arrangement provided as part of a wider package of measures (see also Guest 2002).

2   Other participants were slightly more colourful in their observations here, suggesting that 'Commuting's a bloody mug's game. Don't get me wrong, I've done it myself for an hour and half a day, whatever. It's still bloody stupid and a waste of time.' CEO, female with two children, IT MNC, UK SE region.

3   As detailed in Chapter 4, this IT employer dataset is composed of 150 firms employing 8,068 workers in the Dublin/Cambridge regions (average 19.3% female workforce). These are predominantly small firms (50 employees or less), 90% with annual revenues up to £5 million, and with operations established locally in Dublin/Cambridge from the 1990s onwards. The skewedness of this IT firm sample towards SMEs is consistent with the larger populations of IT firms in both regions.

4   Some managers also commented on the need to ensure that workers' use of flextime and homeworking in particular did not exacerbate problems of time- and strain-based work-life conflict by simply displacing it into workers' homes (that is by reinforcing a 'work anytime, anywhere, anyplace' mindset).

5   Following Gibson-Graham (2006), the potential dangers of using the terms 'alternative' or 'non-standard' to describe working arrangements are that it 'subordinates what it designates to the "mainstream" … [and] affirms the dominant by identifying the deviant' (p. xxii). Usefully, however, it also signals that the status quo is problematic, and that there is a push to change things, to pose a challenge to the 'mainstream' work norm.

6   This tripartite frequency/formalisation/structure communication typology draws on ideas by Hoegl and Gemuenden (2001), to which they also add a fourth dimension, 'openness' of exchange (degree of non/withholding important information).

7   While the majority of the 2010 IT worker survey participants were UK-based, these workers exhibited a similar skills profile to the pre-recession IT worker survey sample with an average 11.6 years' work experience in the IT sector across 3.8 different employers. Additionally, 39% of the 2010 IT worker survey cohort (N = 147) had a Masters degree or higher; 70% were employed in technical managerial, technical or research roles at the time of the survey; and only 7% in HR, marketing or sales roles. The 2008 and 2010 IT worker survey samples also exhibit similar age distributions and proportions of working mothers with dependent children of pre-school age and in full-time primary education.

8   Recessionary changes in employer attitudes to work-life provision are evidenced through a comparison of the UK's Workplace Employee Relations Survey (WERS) in 2004 and 2011, evidencing an increase in the majority managerial expectation that responsibilities for managing work-life conflict rest with workers (from 66% to 77%).

Chapter Seven
Work-Life Balance, Cross-Firm Worker Mobility and Gendered Knowledge Spillovers

Introduction

> It is misleading to think of scientific breakthroughs as disembodied information which, once discovered, is transmitted by a contagion-like process in which the identities of the people involved are largely irrelevant. (Zucker and Darby 1996: 12709)

This book is concerned with the everyday work-lives of the engineers, scientists and technologists whose collective labours are ultimately responsible for (re)producing and sustaining some of the world's most high-profile high-tech regional economies. The everyday hardships experienced by these knowledge workers and their families as they struggle to reconcile competing responsibilities of paid work, home and family formed the focus of Chapter 5. Much more than just a problem for workers, Chapter 6 showed how those work-life conflicts also have major consequences for the firms that employ them, powerfully shaping their intra-firm capacities for learning, innovation and socio-economic competitiveness. In this chapter, I extend that *intra*-firm analysis to explore the role of work-life conflict and WLB provision by high-tech employers in also shaping networks of learning and knowledge exchange *between* firms.

As a cornerstone of the regional learning agenda, economic geographers have long identified the role of mobile skilled labour in accelerating the movement of embodied knowledge, expertise and technological capabilities between firms, and thereby enabling firms to adapt to changing market

Work-Life Advantage: Sustaining Regional Learning and Innovation, First Edition. Al James.
© 2018 John Wiley & Sons Ltd. Published 2018 by John Wiley & Sons Ltd.

conditions and avoid technological lock-in. Paradoxically, however, despite this concern with high-skilled 'labour', the regional learning and innovation agenda has offered almost zero engagement with the wider *labour geographies* agenda, concerned 'to see the making of the economic geography of capitalism through the eyes of labour' (Herod 1997: 3). Consequently, geographers have said very little about how workers' multiple identities, varied responsibilities of care and personal life interests beyond the workplace unavoidably shape their abilities to act as agents of knowledge exchange in practice. This in turn has obscured a darker side to knowledge spillovers in regional economies, and reinforced the powerful (yet problematic) economistic claim that cross-firm labour mobility is always and everywhere a good thing. This chapter seeks to address these analytical shortcomings through new engagements with female worker agency, gendered labour geographies of work-life balance and social reproductive care.

Building on the critiques developed in Chapter 2, the first half of the chapter shows how the majority of regional learning insights to date around embodied knowledge spillovers have come exclusively from examinations of the mobility patterns of male workers – or else from analyses that collapse workers into a genderless, unitary category of mobile 'labour' who simply 'switch from one employer to another' (Breschi and Lissoni 2009: 445) to enable improved innovation outcomes. Through new engagements with female technologists, my analysis shows how the embodied 'knowledge externalities' so widely celebrated in regional learning and innovation studies – and in regional development policy – are in fact also premised on gendered work-life conflict, unequal divisions of household reproductive labour, uneven and often inadequate WLB provision by employers, and worker concerns that stretch far beyond the technological to include family, care and quality of life. As such, WLB provision is shown to have significant benefits for IT firms' abilities to access cross-firm knowledge spillovers relative to firms with lower levels of provision.

Crucially, I also show how work-life concerns also *constrain* cross-firm job-to-job mobility and embodied knowledge spillovers, as some technology workers deliberately seek to avoid the potentially disruptive impacts of a new job on delicately balanced commuting patterns combining nursery, school and a partner's commute, redistribution of domestic duties, possible relocation of home and school, and established networks of family and community support. In so doing I reveal a darker side to knowledge spillovers. The final part of Chapter 7 explores the role of urban carescapes and national welfare regimes in differently shaping the connections between work-life balance and knowledge spillovers. It also identifies the significance of 'luxury' concerns around work-life balance in *continuing* to shape patterns of job-to-job mobility by female technology workers in the aftermath of economic crisis. In these ways, my analysis suggests that we need to unlearn a few more 'universal' (masculinist) tenets of economic theory at the heart of the regional learning agenda.

Embodied Knowledge Spillovers and Regional Competitive Advantage: The Story So Far

In the wake of the widely purported shift to the 'knowledge economy', the capacities of firms and regions to foster interactive processes of learning and innovation are now widely recognised as fundamental to their sustainable economic advantage. The holy grail of the regional learning agenda, then, has been to demystify the mechanisms and agents through which 'knowledge generated within innovative firms and/or universities is somehow transmitted to other firms' (Breschi and Lissoni 2001: 980) in support of enhanced economic performance.

One widely identified mechanism is the physical mobility of labour – or 'learning-by-hiring' (Song et al. 2003) – in which 'departing workers ... are carriers of vital information and experiences that follow them to their next workplace' (Boschma et al. 2009: 171). Cross-firm labour mobility enhances firms' innovative capacities because it allows work teams to bring new constellations of ideas, skills and accumulated experience to bear upon novel problems, and increases the potential for unexpected ideas, new interpretations and learning synergies as colleagues expose each other to alternative viewpoints, epistemic habits and critiques (Östbring and Lindgren 2013). And much more than a once-and-for-all knowledge transfer, mobile workers may also maintain advantageous *ongoing* knowledge links between their new and previous employer via personal networks, further integrating these firms into larger communities of practice (Boschma et al. 2009). In this way, the physical mobility of labour between firms is widely heralded as playing an important role in determining localised patterns of innovation and growth (McCann and Simonen 2005), enabling firms to adapt more effectively to changing market conditions and avoid technological lock-in (Bathelt et al. 2004), and understanding the benefits of spatial colocation for the competitiveness of firms and regions. Indeed, for some scholars, labour mobility is identified as *the* main mechanism of local knowledge spillovers, superior to formal networks of cross-firm collaboration and informal social ties (Breschi and Lissoni 2009).

Responsibility for advancing our understanding of regional knowledge spillovers through mobile labour has lain largely with an influential group of founding authors and disseminator studies, with the most highly cited studies listed in Table 7.1. A more recent set of analyses is listed in Table 7.2, showing how this research agenda has been extended over time. Notable among these studies is AnnaLee Saxenian's seminal (1994) account of young, mobile, male engineers in Silicon Valley who 'shifted between firms so frequently that mobility not only was socially acceptable; it became the norm' (p. 34), with minimal disruption to personal, social or professional ties (p. 35). Early research was also concerned to document cross-firm knowledge flows empirically using patents and citations as a 'paper trail' (see, e.g., Jaffe et al. 1993), later extended to the physical mobility of patent holders themselves (Almeida and Kogut 1999). Similarly, Zucker and

Table 7.1 Analysing regional knowledge spillovers I: influential/founding studies.

Author/study	Title	Evidence base/sample/method	Worker label	Disaggregation by gender?	Citations 6 March 2017 Google Scholar
Saxenian 1994. Harvard University Press	*Regional Advantage: Culture and Competition in Silicon Valley and Route 128*	143 in-depth interviews (1988–1991) with entrepreneurs, industry leaders, corporate executives and representatives of local business associations, government and universities in Silicon Valley and Route 128. Only 15 research participants were women.	'engineers', 'computer professionals', 'technical professionals'	NO	11,075
Jaffe et al. 1993. *Quarterly Journal of Economics* 108(3): 577–598	Geographic localization of knowledge spillovers as evidenced by patent citations	Paper trail of knowledge spillovers evidenced by 2 cohorts of US 'originating' patents granted to US universities and firms. 1975 cohort: 950 patents with 4750 citations by 1989 end. 1980 cohort: 1450 patents with 5200 citations by 1989 end.	'inventors'	NO	7,390
Almeida and Kogut 1999. *Management Science* 45(7): 905–917	Localization of knowledge and the mobility of engineers in regional networks	Database of career paths (1974–1994) of 438 individuals holding major semiconductor patents; 12 regions of significant semiconductor activity in the USA; 303 semiconductor patents (filed in 1980 and 1985). 174 intra-regional moves, 181 inter-regional moves.	'engineers' (p. 905), 'patent holders' (p. 913)	NO	2,482

Source	Method	Concept	Body counted?	N
Zucker and Darby 1996. *Proceedings of the National Academy of Sciences* 93: 12709–12716	Sept 1990 release of GenBank documenting all scientific journal articles reporting on genetic sequencing up to April 1990. 4061 articles which identify 327 star bioscientists worldwide (and 6082 distinct collaborators) and their employing universities, firms, research institutes and hospitals.	'star bioscientists' (p. 12709), 'intellectual human capital' (p. 12710)	NO	712
Fallick et al. 2006. *Review of Economics and Statistics* 88(3): 472–481	US Current Population Survey (1994–2001) data on month-to-month mobility of 44,202 individuals (samples males only, 4+ yrs college education and living in US metro areas with IT clusters).	'human capital externalities' (p. 473)	NO (females deliberately excluded)	420
Henry and Pinch 2000. *Geoforum* 31: 191–208	50 in-depth interviews (managers, designers, engineers); career mapping of 100 leading design engineers.	Embodiment of knowledge 'in a thinking, breathing body such as the engineer' (p. 195)	NO	299

Table 7.2 Analysing regional knowledge spillovers II: extending the debate in economic geography.

Author/study	Title	Evidence base/sample/method	Worker label	Disaggregation by gender?
Agrawal et al. 2006. *Journal of Economic Geography* 6: 571–591	Gone but not forgotten: knowledge flows, labor mobility, and enduring social relationships	US Patent and Trademark Office: North American sample of mover inventors documented in 59,734 cited patents (1990) plus 57,878 cited patents (1989).	'movers' (p. 580), 'inventors of focal patents' (p. 583)	NO
Bienkowska et al. 2011. *Geografiska Annaler B* 93(1): 21–39	Brain circulation and flexible adjustment: labour mobility as a cluster advantage	Mixed methods. Labour mobility survey data on all individuals employed in Kista and Mjärdevi ICT clusters, Sweden 1997–2002 (18,501 ICT workers Kista and 3403 ICT workers Mjärdevi). Plus 30 interviews with firm representatives in both regions 2005–2006.	'brain circulation mobility' (p. 24), 'churning of skilled labour' (p. 34)	NO
Breschi and Lissoni 2009. *Journal of Economic Geography* 9: 439–468	Mobility of skilled workers and co-invention networks: an anatomy of localized knowledge flows	Subset of European Patent Office (EPO) dataset 1978–2002 targeting all patent applications with at least one US inventor and applied for by a US organisation in organic chemistry, pharmaceuticals or biotechnology (66,349 patent applications signed by 63,188 distinct inventors across 5820 different organisations). Logistic regression.	'mobile inventors' (p. 439), 'mobility of technologists' (p. 465)	NO
Boschma et al. 2009. *Journal of Economic Geography* 9: 169–190	How does labour mobility affect the performance of plants?	Swedish national economy: analysis of 101,093 job moves, ASTRID longitudinal micro-database containing info on all Swedish inhabitants, firms and workplaces (2001–2003). Logistic regression.	'skill portfolios' (p. 169), 'human capital' (p. 174), 'competence portfolios' (p. 182)	NO

Reference	Title	Data and methods	Terminology	
Eriksson and Lindgren 2009. *Journal of Economic Geography* 9: 33–53	Localized mobility clusters: impacts of labour market externalities on firm performance	Swedish national economy: 150,000 individuals changing jobs between 50,000 workplaces (2001); via ASTRID micro-database containing annual information on all Swedish inhabitants, firms and workplaces (1985–2003). Logistic regression.	'individuals changing jobs' (p. 38), 'job changers' (p. 39), 'flows of embedded knowledge' (p. 48)	NO (job changers) (model includes % female workforce at firm level)
Power and Lundmark 2004. *Urban Studies* 41(5/6): 1025–1044	Working through knowledge pools: labour market dynamics, the transference of knowledge and ideas, and industrial clusters.	Time series dataset 1.1 million workers in the Stockholm labour market (1990–1995); labour mobility analysis within the Kista ICT cluster cf. the rest of Stockholm LLM.	'labour mobility'	NO
Trippl 2013. *Regional Studies* 47(10): 1653–1667	Scientific mobility and knowledge transfer at the interregional and intraregional level	Mobility survey 2,481 'star scientists' identified through ISI Highly Cited Papers database (papers by 5,600 of world's most highly cited scientists, 1981–2002). 92.6% male, 5.6% female (1.8% missing).	'star scientists'	PARTIAL

Darby (1996) identified the crucial 'boundary spanning' role played by the movement of star scientists' 'embodied intellectual capital' between universities, firms, research institutes and hospitals, arguing that 'technology transfer is about people' (p. 12715). And in Henry and Pinch's highly cited analyses of Oxford's Motor Sport Valley, the embodied 'churning of knowledge' across firms is tracked through the rapid and continual movement of designers, engineers, managers and drivers; high rates of firm death rates and new firm formation in which key personnel are 'recycled'; and interlocking company directorships (Henry and Pinch 2000). This mapping revealed a move on average of once every 3.7 years, and a total of eight moves in an average career, 'helping to raise the knowledge throughout the industry' (Henry and Pinch 2000: 198).

Extending these mobility/learning insights, more recent analyses have also examined different degrees of complementarity between the knowledge that new employees embody and the existing knowledge base of the receiving firm (e.g. Boschma et al. 2009; Östbring and Lindgren 2013). Scholars have also continued to measure the quality of mobile 'human capital' and 'competence portfolios' through common proxies including educational attainment, number of previous employers, patent citations, journal citations and industrial tenure. In combination, then, this body of research offers some important advances over earlier esoteric notions of knowledge and innovation residing 'in the air' (Marshall 1890) or in the 'buzz of urban life' (Storper and Venables 2002), by demonstrating that cross-firm diffusion of tacit knowledge is not some disembodied, abstract process. Rather, it is always and everywhere *peopled*.

Zero Drag Labour Mobility as Regional Learning Advantage? The Limits of Previous Analyses

At the heart of the regional learning and innovation literature, then, labour mobility is widely understood to have positive effects on the performance of firms in clusters (Eriksson and Lindgren 2009) because it serves as a mechanism for acquisition of external knowledge that helps firms to avoid lock-in (Boschma et al. 2009); erodes differential knowledge advantages between firms (McCann and Simonen 2005); and offers a way for firms to access knowledge developed at other firms without their approval (Song et al. 2003). So far so good. More problematic, however, is the widespread assumption that underpinning these learning advantages, skilled labour is motivated to move between firms primarily as a function of technological imperatives, such that 'constraints to mobility are often regarded as market imperfections' (Fischer and Malmberg 2001: 358).[1] Likewise, commentators have noted that:

> The transfer of tacit, high value knowledge between actors is held to be a benign process … related to the rational actions of the workers as they seek to solve technical problems and thereby buttress the viability of their firms. (Wilson and Spoehr 2010: 42–43)

In this section I seek to move beyond this 'benign technical rationality' lens, highlighting three peculiar biases which currently constrain geographical analyses of the cross-firm mobility of 'labour' and its implications for regional knowledge spillovers. In many ways, these epitomise the wider limits to the regional learning and innovation agenda as articulated in Chapter 2, in which the majority of 'zero drag' analyses to date remain firm-centric, gender-blind and/or rooted in the masculinist myth of the disembodied ideal worker for whom work is primary and the demands of family and personal life deemed insignificant. Same old, same old...

Dehumanised, firm-centric conceptions of 'mobile labour' as knowledge commodity

Despite the apparent prominence of 'labour' as a key focus of regional learning analysis, paradoxically workers themselves remain theoretically invisible, treated 'not as social agents capable of making landscapes in their own right but, rather, as simply an aspect of capital' (Herod 2009: 22). Mobile labour enters these analyses merely as embodied sets of transferrable competencies, which in turn have the potential to bolster knowledge exchange and production processes in the firms to which they move. In contrast, the lived experiences of the workers in accumulating these skills and (re)producing these 'knowledge spillovers' are treated as at best unremarkable, or worse analytically irrelevant.

This dominant approach to labour is evident in a plethora of instrumentalist capital-centric terms used in the regional learning literature that variously reduce workers to 'competence portfolios' (Breschi and Lissoni 2001: 993) or 'carriers of critical knowledge flows' (p. 991); 'stocks of human capital' (Lawton-Smith and Waters 2005); 'human capital externalities' (Fallick et al. 2006); 'regional knowledge assets' (Faggian and McCann 2006); 'knowledge capital' (Jaffe et al. 1993); or even just disembodied 'mobile brains' (Bienkowska et al. 2011)! Despite this diversity, these terms are united in their encapsulating a particular approach to mobile labour that is dehumanising, and which reduces people to their 'knowledge diffusing function' (Breschi and Lissoni 2001: 991).

In some ways, this instrumentalist view of mobile (and male) labour harks back to Marshall's original 'triad of localisation externalities' that accrue to firms as they collocate spatially:

> A localised industry gains a great advantage from the fact that it offers a constant market for skill. Employers are apt to resort to any place where they are likely to find a good choice of workers with the special skill they require; while men seeking employment naturally go to places where there are many employers who need such skills as theirs and where therefore it is likely to find a good market. (Marshall [1890] 2013: 225–226)

While subsequent work has sought to document more dynamic knowledge externality effects of labour market mobility beyond Marshall's 'constant market for skill', we still know very little about workers' everyday lived experiences of (re) producing cross-firm knowledge spillovers in practice – nor of the geographical possibilities for them to effect improvements in their work-lives from doing so. In addition, regional learning accounts of the apparently straightforward cross-firm spillovers of embodied knowledge and skills are generally silent about workers' *differential* abilities to identify new employment opportunities and then success-fully to negotiate employment in other firms. Indeed, regional learning analyses say almost nothing about technology workers who are *not* mobile, but who instead stay put.

Cross-firm knowledge mobilities and the masculinist universal (or, male = 1)

A second major blindspot within accounts of cross-firm knowledge spillovers through the physical mobility of labour concerns the widespread analytical invis-ibility of female knowledge workers in (re)making these regional geographies of learning and innovation in practice. Instead, the majority of insights around knowledge spillovers to date have come from examinations of male worker mobility, or else from analyses that collapse workers into a genderless, unitary category of 'mobile labour'. These problems are illustrated by a number of well-known and highly cited studies, as evidenced in Tables 7.1 and 7.2.

On one level, these patterns emerge from the numerical dominance of men in the industries which form the objects of study. For example, in Henry and Pinch's (2000) analysis of the 'churning' of 100 leading designers, managers and engi-neers between firms in Oxford's motor sport industry, all of the career history biographies tracked are male. Similarly, in AnnaLee Saxenian's (1994) account of labour mobility in Silicon Valley's semiconductor industry, only 15 of her 143 research participants were female. However, in other studies based on much larger and/or multi-industry datasets with higher proportions of female workers, regional scholars have also failed to disaggregate by gender. Based on an impres-sive time-series dataset of 1.1 million Swedish workers, Power and Lundmark (2004) explore intensities of intra- and inter-firm mobility of professional workers in Stockholm's prominent ICT cluster (centred on Kista Science Park). However, whilst acknowledging that women comprise 29.3% of ICT workers in this cluster, Power and Lundmark fall short of comparing male and female labour mobility patterns, instead lumping both groups together in an apparently genderless mass. A similar tendency is apparent amongst other Swedish labour mobility studies using the ASTRID database, which includes detailed information about all inhabitants, firms and workplaces at the national scale. Despite this wealth of data, analyses by Boschma et al. (2009) and Eriksson and Lindgren (2009) of the effects of skilled labour mobility on firm performance in Sweden fall short of distinguishing workers by gender.

The same gender myopia is also evident in influential studies that track knowledge mobility through patents. For example, while Zucker and Darby (1996) make clear that 'the identities of the people involved' matter in technology transfer, they fail to disaggregate their consequently faceless and genderless database of 337 star scientists on this basis. Likewise Almeida and Kogut's (1999) analysis of regional variations in knowledge spillovers tracked through the career paths of 438 (apparently genderless) engineers with patent holder records in the US semiconductor industry. And while Agrawal et al.'s (2006) analysis of North American embodied knowledge flows makes clear that they 'identify "movers" in these data by examining the inventor names ... including the middle initial if present' (p. 576), again there is no attempt to disaggregate inventors by gender within this large dataset of 62,817 named movers.

It is also interesting to note the variety of methodological strategies through which these various female exclusions emerge. These include the indirect exclusion of many female workers on fractional contracts on the basis that 'the analysis should focus on *full-time working individuals* who have established themselves on the labour market' (Eriksson and Lindgren 2009: 39, emphasis added).[2] In other studies the exclusion of 'confounding' female workers from cross-firm job-to-job mobility analyses is much more explicit and direct:

> We restrict our sample to those having a minimum of four years of college who also live in metropolitan areas having information technology clusters. In addition, we focus our analysis on men to eliminate the potentially confounding effect of gender on mobility. (Fallick et al. 2006: 476)

Problematically, these are far from isolated examples. In addition to the studies in Tables 7.1 and 7.2, other regional learning mobility studies that fail to explore the gendering of these localised 'knowledge externalities' include Keeble et al. (1999), Lawton-Smith and Waters (2005), Agrawal et al. (2006) and Östbring and Lindgren (2013). And while a select few studies *do* include gender as a relevant variable in their logistic regression analyses of regional knowledge spillovers through cross-firm worker mobility (Dahl 2002; Thulin 2009), this is typically in the form of a female 'dummy variable' that reduces female technology workers to that which they are not. This tendency within the regional learning literature is consistent with hierarchical dualisms identified by Faulkner (2000) more widely in studies of gender and technology 'in which the masculine side is something and the "other", feminine side defined as the negation of that thing' (p. 782). Or what Spender (1980) refers to as 'minus male'. Consequently, these studies offer little insight into the complex gendering and sexing practices through which 'men and women *supply* their labour on different terms' (Marshall 1994: 44), and which unavoidably shape the operation and outcomes of cross-firm job-to-job (im)mobility.

False abstraction of embodied 'knowledge externalities' from networks of social reproduction

Closely linked to the analytical priority afforded to firms, and general analytical silence around female worker (im)mobility, a third blindspot in regional learning analyses concerns the false abstraction of labour and 'embodied knowledge spillovers' from the networks of household labour and social reproduction which simultaneously reproduce and constrain them. Overwhelmingly, the cross-firm job-to-job mobility of labour is understood in the regional learning literature as 'easy and rapid movement' (Fallick et al. 2004: 2) carried out by individual economic agents who 'switch from one employer to another' (Breschi and Lissoni 2009: 445) to provide 'a steady stream of incoming labour' (Power and Lundmark 2004: 1027) in pursuit of improved technological innovation outcomes (Saxenian 1994). Indeed, within the regional learning literature, 'constraints to mobility are often regarded as market imperfections' (Fischer and Malmberg 2001: 358).

On one level, these conceptualisations are consistent with other positive assessments of high rates of inter-organisational mobility amongst technology workers, reflected in short organisational tenures, and the development of so-called 'boundaryless' portfolio careers (Arthur and Rousseau 1996) which span multiple employers, and thereby eschew earlier models of continuous hierarchical career development within a single company. However, to conceive of workers and their (apparently unimpeded) patterns of cross-firm job-to-job mobility simply in relation to firms' innovative capacities and the needs of 'the market' fails to recognise the myriad ways in which

> Individuals conduct their lives in multiple spheres, including for example: the household, the workplace ... Importantly, people often develop different values and logics relative to these different spheres [and] ... thus carry with them multiple rationalities as they move across spheres, and the rationalities themselves become intertwined. What is commonly understood as *ir*rational is a matter of the operation of *a* rationality, possibly developed in another sphere. (Ettlinger 2003: 153)

In short, the widely documented 'carriers of critical knowledge flows' celebrated within the regional learning literature are also wives and husbands; 'embodied competence portfolios' are also mothers and fathers; 'mobile brains' also daughters and sons; and embodied 'stocks of human capital' also members of household units, extended families, communities and friendship groups. Accordingly, the everyday *abilities* and *willingness* of technology workers to move between firms is far from a unidimensional phenomenon carried out by atomised units of labour driven only by corporate logics to maximise innovative capacity and quarterly revenues on behalf of firms. Reinforcing these problems, the predominance of logistic regressions and large datasets within the regional learning labour mobility literature means that studies typically only document the patterns of cross-firm job-to-job mobility that *did* happen. Workers who remain with their employer drop out of these analyses. These analyses also say nothing about the competing priorities, normative concerns, strategic work-life trade-offs and household and family decision-making processes which underpin observed patterns of cross-firm mobility *and immobility*, nor of how patterns of cross-firm (im)mobility for individual workers change over time in relation to key lifecourse events and stubborn gendered structures of career constraint.

Encouragingly, some regional learning scholars have begun to acknowledge the role of workers' sex and family situation in shaping observed patterns of cross-firm job-to-job mobility (e.g. Dahl 2004; Gray and James 2007; Bienkowska et al. 2011); the difficulties for dual-income households in finding employment for two persons within the same local labour market (Power and Lundmark 2004: 1027–1028); and hence the need for studies of knowledge spillovers to incorporate 'more information about the working-life history of the labour ... and how this is related to plant performance and regional growth' (Boschma et al. 2014: 1686). But in the main, the profound implications of workers' simultaneous embeddedness in extra-firm networks of gendered household relations, social reproduction and care for cross-firm spillovers of embodied knowledge and learning remain heavily under-researched.

Gendered Labour Market (Im)mobility: Insights from Beyond the Regional Learning Boys' Club

In seeking to recentre these regional learning labour mobility analyses, it is imperative that scholars engage with an established body of scholarship in sociology, labour studies and feminist geography that has *already* documented multiple gendered constraints on cross-firm worker mobility. From the mid-1970s onwards, these studies have repeatedly identified the role of workers' gender, marital status, numbers and ages of dependent children, and/or partner's income in differently shaping patterns of cross-firm job-to-job mobility and female career advancement (e.g. Tyree and Treas 1974; Markham et al. 1983; Payne and Abbott 1990). Within this framework, studies have highlighted the difficulties for many women of taking jobs whose working hours conflict with their majority responsibility for childcare (Felmlee 1982; Valcour and Tolbert 2003); constrained patterns of cross-firm job-to-job mobility amongst men and women in dual-earner households (Bailey et al. 2004); and incidences of 'trailing spouse syndrome' in which (predominantly) female workers exhibit compromised patterns of job-to-job mobility relative to male colleagues in order to prioritise their partner's career advancement (Han and Moen 1999; Hardhill 2002). Scholars have also found significant differences in the levels of cross-firm job-hopping between female workers with children and their female colleagues with minor or no home and childcare responsibilities (Dumelow et al. 2000), as the former attempt to minimise the disruptive effect that changing jobs can have on the entire family unit (see Folbre 1994).

Thus, rather than conceptualise workers as discrete, atomised agents, workers are instead conceptualised as members of 'household networks of linked lives' that span work, home and family (Bailey et al. 2004; Elder 1994). As part of these networks of linked lives, childcare provision may involve local resources including friends, grandparents and extended family, and 'mothering networks' of friends and colleagues (Bailey et al. 2004). Accordingly, the age of dependent children and the presence of dependent elders are also identified as significant factors in shaping patterns of cross-firm worker (im)mobility (e.g. Fischer and Malmberg 2001), with more children in the household meaning fewer windows of opportunity when it is deemed possible to make a move, to avoid the disruptive impact on children's education (Green 1997). Also in contrast to notions of atomised worker mobility prevalent in the regional learning literature, research suggests that the additional financial responsibilities of having children tend to *increase* men's attachment to their employer, and to decrease their propensity to change employers (Blau et al. 2002). This alongside additional constraints on the frequency and spatial patterns of cross-firm job-to-job mobility documented amongst dual-professional career couples (e.g. Ackers 2004; Pixley 2008). And divergent from the seemingly time-invariant mobility of 'human capital' within the regional learning literature, studies have also documented the effects of major life events including marriage, birth of children and divorce in differently prompting and constraining patterns of labour market mobility over the lifecourse. This work suggests that cross-firm mobility is generally more feasible for workers at younger ages because of their relative lack of commitments to current employers, households and established social networks (Fischer and Malmberg 2001). However, the implications of these gendered patterns of worker mobility for the dynamics of cross-firm knowledge spillovers remain largely unexplored in these studies.

Extending these insights through widely cited work within feminist geography, Hanson and Pratt have explored the ways in which family–household responsibilities structure women's labourforce participation and negotiation of cross-firm (im)mobility in urban labour markets. Based on a case study of women in the Worcester, Massachusetts metropolitan area,[3] Hanson and Pratt (1995) show how, as a function of female majority responsibilities for everyday tasks of household care, employment opportunities closer to home have greater significance for women than men. They demonstrate the significant role that factors such as job hours, flexibility of hours, proximity to home, spatial proximity to children's daycare, fit with partner's job, fit with children's school schedules and gendered power relations within the household play in constraining women's patterns of job search and job-to-job mobility, and their variations over the lifecourse (e.g. Hanson and Pratt 1991).[4] Indeed, the very spatial positioning of sites of social reproduction (homes, schools, daycare, creche facilities, shops, sports facilities) relative to sites of production means that gender relations of care become inscribed in the very spatial structure of cities (England 1991; Jarvis 2005). Hanson and Pratt (1991) also highlight the role of 'sequential scheduling strategies' of the second and third shifts within dual-career households (see Hochschild 1997), in which couples arrange their paid employment so that one adult is at home at all times in order to care for the children, with consequent implications for both the type of work chosen and its spatial location. In short, they emphasise the dynamic interdependencies between work and home for shaping worker decisions and opportunity structures around cross-firm mobility, and 'the futility of trying to study one in isolation from the other' (Hanson and Pratt 1988: 299).

This latter theme is also echoed in a third body of work that explores the impacts on labour turnover of employer-provided WLB arrangements intended to help workers reconcile competing responsibilities and commitments of work, home and family. Here, studies have documented how employees who have difficulty in spending time with their family or keeping social commitments are more likely to be emotionally exhausted, which in turn impacts negatively on their job satisfaction and influences their intention to leave a particular company (Karatepe and Uludag 2007). However, the dominant focus within these studies tends to be on the effect of work-life conflict and/or employer-provided WLB arrangements on workers' turnover/quit intentions (e.g. Good et al. 1996; Allen et al. 2000; Batt and Valcour 2003; Boyer et al. 2003), and/or the associated costs to firms of replacing workers who leave (e.g. Forsyth and Polzer-Debruyne 2007).[5] In contrast, the subsequent workplace destinations of these workers once they have left a particular firm are undocumented. Likewise, we know surprisingly little about the work-life advantages that accrue to these workers, and the learning advantages that accrue to their employers, as a function of that cross-firm job-to-job mobility.

Recentering the Regional Learning and Innovation Research Agenda

Despite the plethora of sociological, labour and feminist studies that have documented multiple gendered constraints on cross-firm worker mobility, there remains an underwhelming lack of engagement with these analyses within the expansive regional learning and innovation agenda. The result is that the regional learning literature fails to recognise the labour geographies that are fundamental to its core objects of study, and reinforces the powerful (and problematic) economistic claim that cross-firm labour mobility is always and everywhere a good thing. In response, the alternative regional learning analysis developed in the second part of this chapter extends this 'intellectual trading zone' (Barnes and Sheppard 2010). It explores the role of gendered everyday struggles by 'labour' to reconcile activities of home, work and family, and of different levels of work-life provision by high-tech employers in collectively shaping the cross-firm (im)mobility of workers; and hence firms' abilities to attract the knowledge, skills and competencies which they embody.

Why do 'competence portfolios' need work-life support? A quick recap

As explored in Chapter 5, everyday experiences of work-life conflict in the UK and Irish IT sector are less a function of ongoing long hours *per se* than of the temporal variability of work hours over the course of software product development and sales life cycles. Other commonly identified sources of work-life conflict include the need for rapid response to client system emergencies, communication between international work team members in different time zones, being on-call for troubleshooting, international customers and weekend travel, and the continual need to maintain skill-sets in the context of a constantly changing industry. Outside of work, in addition to domestic labour and childcare responsibilities, Irish and UK IT workers were also shown to maintain a wide range of other personal responsibilities and interests including self-building a new house, running a Cub Scout pack, singing in a choir, doing voluntary work, regularly taking time out for international travel, labour organising, caring for horses and home schooling children. Add to these the daily demands of IT work, and the everyday stresses and strains of juggling work, home and personal life in the Irish and UK IT sector become clearer, with the most commonly cited outcomes identified as missing out on children's activities, interrupted sleep patterns, stress and exhaustion impacting on relationships with children and partners, working when feeling unwell, missing out on leisure time and hobbies, and an overall reduced quality of life. And whilst there is no simple homogeneity, Chapter 5 showed that experiences of work-life conflict in the IT sector are particularly acute for female professionals with young children – many of whom, despite a gender convergence in parents' contribution to childcare time, continue to undertake the majority share of household labour and childcare responsibilities.

In response to these challenges, research participants identified their preferred employer-provided policies and practices – those they regarded as most effective in helping them to reconcile everyday work-life conflicts. Chapter 5 showed that there is no panacea: work-life requirements vary not only within gender groups (by job function, department, household situation) but also for individual workers over the lifecourse. Thus, while working mothers and fathers with young families often identified extra-statutory provision of maternity and paternity leave, plus employer-subsidised childcare, as their preferred employer-provided WLB arrangement, this was less important for working parents with children of school age (some of whom preferred term-time working), and an irrelevance for research participants without children. And while other research participants favoured flextime, other workers with lengthy commutes instead preferred greater spatial flexibility of work, citing the personal advantages gained from working from home and of 3- or 4-day work weeks as: more time for extra-curricular activities, hobbies and interests; better quality of personal time; reduced levels of stress and exhaustion; increased autonomy; and, particularly for working parents with young families, better quality of time spent with children. In short, different home and work factors mean work-life provisions suitable for one cohort of employees at one point in time may have little or no effect in reducing work-life conflict for another.

In seeking to integrate the productionist focus of the regional learning and innovation literature with concerns around social reproduction, family well-being and quality of life, this diversity of workers' WLB preferences raises important questions. Specifically, how do these preferred arrangements compare with the kinds of work-life policies and practices that different employers are making available to workers? And hence, what are the consequences of those (mis)matches for shaping the kinds of cross-firm labour mobility pathways widely identified as enabling cross-firm knowledge spillovers in dynamic regional economies?

Uneven work-life provision and its consequences for knowledge spillovers

The IT employer survey reveals an unevenness in the specific *types* of formal WLB arrangements employers are willing to make available to employees (Table 7.3). The most widely available employer-provided work-life arrangements are flextime (67% of firms); working from home one or two days a week (66% of firms); and part-time working (54% of firms). Compare this with the much greater diversity of workers' WLB preferences identified in Chapter 5 which include, for example, the preferences of working mothers and fathers with young families for extra-statutory provision of maternity leave (available in only 22% of firms), extra-statutory paternity leave (available in only 11% of firms) and employer-subsidised childcare (available in only 6% of firms). Likewise, identified worker preferences for compressed working weeks (available in only 37% of firms), or term-time working (available in only 9% of firms). The employer survey data also document unevenness in the *total suites of WLB provision* available to workers across different IT employers (Figure 7.1).[6]

Against this backdrop of uneven provision of WLB arrangements by IT employers, the analysis suggests that work-life balance considerations are actively shaping workers' decision-making processes around cross-firm job-to-job mobility, in ways previously unexplored within regional learning and innovation analyses. The interviews with IT workers and managers reveal three qualitative dimensions to this phenomenon, all of which disrupt the dominant masculinist conceptions of worker mobility motivated simply by 'the cause of advancing technology' and 'far greater loyalty to one's craft than to one's company' (Saxenian 1994: 36).

Table 7.3   Dublin and Cambridge IT employer provision of (formal) work-life arrangements (2008).

Category	Formal WLB arrangement	Dublin N = 74	Cambridge N = 76	Combined N = 150
		%	%	%
Flexible work arrangements	**Flextime** (flexible beginning/end time, sometimes with core hours)	**73**	**61**	**67**
	Flexplace (work from home 1 or 2 days a week)	**74**	**58**	**66**
	Flexplace (work from home 3 or 4 days a week)	35	53	44
	Job sharing (one job undertaken by 2 or more persons)	9	4	7
	Annualised hours	8	11	9
Reduced work hours	**Part-time work**	**49**	**59**	**54**
	Compressed work weeks (total work hours in 4 days rather than 5)	31	29	37
	Term-time working	8	9	9
Personal leave	Extra-statutory maternity leave	32	12	22
	Extra-statutory paternity leave	14	9	11
	Career break/sabbatical	19	7	13
Practical help with childcare	Employer-subsidised childcare	4	8	6
	Information referral service for childcare	4	3	3
	Workplace nursery	3	1	2
Other	WLB counselling/training	15	4	9

(i) Voting with your feet: cross-firm mobility in pursuit of better formal WLB provision
Divergent from regional learning portrayals of labour mobility in pursuit of improved technological innovation outcomes, the interviews evidenced workers' concerns to move between IT employers in search of more comprehensive levels of *formal* WLB provision. That is, workers looking to move from employers on the left-hand side of Figure 7.1 to employers situated on the right-hand side by 'voting with their feet', in order to reduce everyday work-life conflict. Participants described the key moments in the hiring process at which work-life requirements were invoked. These included one female CEO in Cambridge who explicitly articulated hers at the outset of her job interview: 'I said to all the VPs in my very first sentence, "and it must honour my commitment to my children." And they looked at me funny, like, okay, that's fine. But it was very important to me that that was something that I was putting in to the mix right up front.' For other participants, work-life considerations were invoked later in the hiring process:

'With my current employer [US MNC], they promote teleworking for work-life balance. I rejected a job offer from a company closer to home because they were not

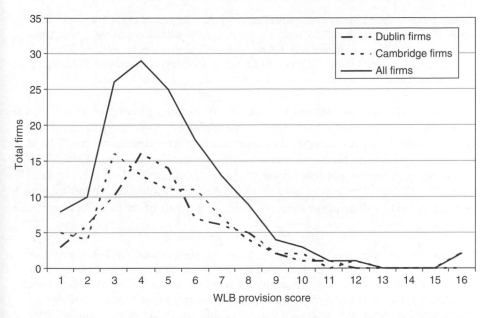

Figure 7.1   Unevenness of total suites of work-life provision across IT employers (Dublin and Cambridge 2008).

open to the idea of working from home or even starting 30 minutes later than the others, to sync my commute with my wife. It is a feature that keeps me in the current company.' Software Engineer, male, IT MNC, Dublin

In addition to these stated intentions by IT workers, the interviews also identified a recognition of this phenomenon by IT employers, its significance for particular workforce cohorts, and their willingness (albeit in some cases under duress!) to honour it. Thus for one HR manager at a large US IT multinational in Dublin, 'we look at what people are looking for when they join the company, and at 35–40 [years old] people are tending to have marriage, children, and the most important thing is flexible working. Compensation doesn't even come on the list of five most important things for women at this particular time in their lives. They're looking at the company and saying "this is not a company where it's good to have career and kids". So they move to a company where they feel it's a good place to come back and have a child and work.' The point, then, is that 'if you're a company that's not attractive to women you've got a big problem' (Human Resources Manager, female, US-owned IT MNC, Dublin).

Similar conclusions were also voiced in other IT firms:

'Employees are saying, "Why should I come and work for you? What sort of flexibility do you have?" I'm helping hire a new department head and we've identified a lady and she says, "I want six weeks' holiday, because I'm a single parent, I've got a son and I like to spend the summer with him." So the company's like, "You won't have time to take six weeks off." "Well, I'll have to reconsider my position then," and she's serious. She will turn a job down if she can't get six weeks' holiday. They *will* give her six weeks' holiday!' Director of Human Resources, male, two children, IT MNC, Dublin

Importantly, the data suggest that this WLB mobility phenomenon is far from limited to women; rather, that there is 'an awful lot more guys taking lifestyle choices around their career: going to companies where they know they'll be able to leave at half four to get home and actually have dinner with their kids. Whereas that in the tech sector that was unheard before' (Marketing Manager, female, US IT MNC, Dublin). Examples here included male technologists willing to incur a pay cut as part of their move between firms in pursuit of an improved work-life balance:

'Firms that are either very pro-family or have a full suite of work-life balance policies … top priorities. My husband changed his job about 15 months ago. He had four offers and he took the one that enabled him to get home for six o'clock to have dinner with us every night, that was his primary driver, he took a three grand pay cut compared to these other offers, on that sole point. He's a hardware engineer.' Sales Manager (previously Engineer), female, two pre-school children, 2-day work week, ICT MNC, Cambridge

The interviews also documented other less typical cross-firm mobility decisions made by some IT professionals in pursuit of an improved work-life balance, including one female software programmer who generally quits her job every two years to 'just have three months off, do something else'. This she explained resulted from 'never having really got over the loss of student holidays', because 'just thirty days' holiday a year, that's really limiting – maybe I could ask for a sabbatical, but you have to work five years to get six months off'. Again, the technological inno-vation motivation that three decades of mainstream regional learning research tells us is driving such cross-firm mobility is difficult to spot here.

In combination, then, these various quotes point to the inseparability of 'agents of innovation' from wider networks of household labour and social reproduction and the major role of work-life balance concerns in shaping cross-firm job-to-job mobility strategies amongst knowledge professionals – and in ways that the regional learning and innovation literature has largely failed to recognise.

(ii) Worker mobility and the search for supportive managers
and cultures of WLB acceptability
Reinforcing workers' concerns around formal WLB provision by IT employers, the interviews also identified how patterns of cross-firm worker mobility are also motivated by perceived differences in *informal* workplace practices that enable and sanction the uptake of formally provided work-life arrangements in practice:

'There was incredibly high turnover at one point, I mean it was just manic. It's calmed down, a lot of people tell me "I really appreciate the work-life balance policies." And it's not just the policies, they appreciate the culture which allows those policies to operate in. It's a big factor for lots of people. I know from other people who work in other companies that it's just not as acceptable to work at home or to come in late one morning, to manage your own time.' Human Resources Manager, female, IT MNC, Dublin

In addition to the acceptance by colleagues, the contents of these workplace 'cultures of acceptability' also include managerial buy-in and senior ratification of non-traditional patterns of working. Indeed for one HR manager in Dublin, 'the biggest thing of all is working with the managers and getting their buy-in as to why we need to do this. I don't think we need a huge amount more policies'. The key role of managers as work-life balance gatekeepers in motivating some workers to move to other employers was also identified by participants in the Cambridge region, highlighting a mismatch between formal WLB provision by firms (*de jure*) and what workers could realistically expect to use in practice (*de facto*). As one female IT specialist explained:

'My experience is to do with your boss, who are you immediately reporting to. In my previous company, I had a very tough time and that's the main reason for me to look elsewhere. I have [two] children and one of them is having a health problem since she was one and a half years old. In addition to that my husband was doing a lot of travelling. So with all the travelling, two kids on my own, it was not possible. HR gave me leave to work from home one day a week, but my boss was very against it. Initially he was very resistant, but then he totally cut that off, he put his foot down.' IT specialist, female, two children (pre-school), SME, Cambridge

How far managers and co-workers are willing to accommodate colleagues with extensive caring commitments was recognised as shaped by their own personal experience (or lack of it) of work-life conflict – 'you know, got the same sort of situation and schedule as the rest of us' (Head of Technology, female, MNC, UK SE region). This rather than 'that's not how it was done in my day and I came through and I turned out all right, so therefore that's how the company runs' (this view as described by one female IT Consultant whose male senior colleagues 'have yet to ride off into the retirement sunset'). The contrast between the following pair of worker experiences is also particularly instructive:

'I'm the only woman in the office, so as far as I can see there isn't the same requirement amongst my colleagues for balance, for juggling. One of the guys doesn't have any kids yet, the other three have stay-at-home wives. So if you don't have somebody who's feeling the same sort of pressure as you, then you're just not in the same boat basically … One guy in my team, he's almost taken it personally at this stage, like "What, you're not available? But I'm scheduling this call now!"' Software Business Development Manager, female, two young children, 3-day work week, MNC, Dublin

'My team, 70% of the people have two or more children and, all of them work from home so teleconferences will often be filled with people appearing in the background, "Where's the shopping, mum?" You get those types of interruptions and it's just accepted … sometimes you have to turn around and say, "Look I have to put you on hold for two minutes, someone's just spilled a glass of milk" or whatever.' Senior Software Developer, male, two young children, IT MNC, Dublin

Indeed, these informal work-life support effects are particularly powerful in the IT industry as a function of decentralised management structures in which autonomous work teams are disciplined by subtle yet strong systems of 'subjective' or 'normative' control through peer pressure (Kunda 2006; Upadhya and Vasavi 2006).[7]

(iii) Work-care patchwork quilts and gendered constraints on cross-firm job-to-job mobility

In addition to different patterns of formal and informal work-life provision across employers, the third qualitative dimension through which WLB considerations are actively shaping workers' decision-making processes around job mobility relate to significant mobility *constraints*. This as a function of the potentially disruptive impacts of any new job on workers' families and households. Here interviews identified workers' concerns to avoid disruptions to delicately balanced commuting patterns combining nursery, school and a partner's commute; redistribution of domestic duties; possible relocation of home and school; and disruptions to established networks of family and community support (i.e. other forms of WLB provision outside of the workplace). These complex work-care commuting trajectories variously combining multiple workplaces, family members, crèche facilities, schools, shops, private care purchases and traffic conditions (both to work and from work) are well illustrated in the following quote:

'I come in early (otherwise a 20-minute drive can be 40 minutes, depends on traffic). But typically if I leave home at half seven I'll be in work for about ten to eight. I finish up here about six o'clock usually. One of the reasons when we bought/live where we do is because it's very close to [the university] because we decided we would pick my husband's workplace as a hub. The three days I work [my husband] drops and collects the children, so one goes to school opposite the university and one goes to crèche on the university, so he drops and collects them. And then the other two days that he travels for work, so he would leave on a Wednesday evening, and then I do the stuff on Thursday and Fridays. So, it's a massive burden. And then the pressure in terms of having to get to a crèche or having to get home to release an au pair, or I have to be somewhere to collect them.' Software Business Development Manager, female, two young children, 3-day work week, MNC, Dublin

Interviews pointed to a variety of delicately balanced 'crazy quilts' (Balbo 1987) that workers weave together to resolve their scheduled and unscheduled childcare needs (and in one case care for older relatives) in the context of traffic, school runs and partner's work over a range of timescales. Thus, in contrast to influential notions of the 'boundaryless career', these delicately balanced work-care arrangements were identified time and time again as powerful boundaries that constrain IT worker decision-making around cross-firm mobility. Likewise how the embodied 'stocks of human capital' and 'competence portfolios' that populate mainstream regional learning analyses might be employed as individuals, but are also members of households and extended families:

> 'We've got kids, we need to get them to playschool/crèche at nine o'clock. If I have to go do that, get back to the house, then get into work, I can start work maybe at ten o'clock. If the company has "everybody has to start by nine", that's not going to work. So, it's important for me to work somewhere that's flexible. The necessary things I have to do for my kids, they have to be met, that's a minimum requirement to getting another job. If that's not accepted then no.' Senior Software Engineer, male, IT MNC, on 1-year career break, two children (1 year and 3 years old), Dublin

Indeed, more than simply limiting *cross-firm* job-to-job mobility, the interviews also documented the role of these same crazy quilts in constraining *internal* mobility, most notably amongst several highly qualified working mothers in professional managerial IT roles who had re-entered the labour market post-maternity leave and moved to a less demanding job role (see also Chapter 5):

> 'It's a question of just juggling. When I did decide to go part-time I did very much have to decide in my own head I was not going to continue to approach my career with the exact same level of zeal. That at least for the short term, the next few years until the little one went to school, I was just gonna take it all down a couple of notches … But I don't think I've planned my career as a woman in a very different way to the way that a man would.' Software Business Development Manager, female, two young children, 3-day work week, MNC, Dublin

Significantly, however – and despite these claims of gender parity – my interviews did not document any similar compromise patterns of downward job-to-job mobility amongst working fathers subsequent to having children.

How widespread is work-life-motivated labour mobility among technology workers?

The interviews identified three qualitative dimensions along which WLB considerations actively inform decision-making amongst IT workers (and their households) around cross-firm job-to-job mobility – and this in ways previously unexplored in the regional learning and innovation literature. Significantly, survey evidence from a larger cohort of IT workers (N = 162) suggests that these are far from isolated cases (see Table 7.4). Amongst the IT worker sample in Dublin and

Table 7.4 IT worker mobility in response to uneven work-life provision by employers (Dublin and Cambridge, N = 162).

Cohort	% movers (no longer with their 1st IT employer)	Average employment tenure (yrs)	WLB arrangements provided by previous employer not useful (%)	WLB PUSH Inadequate WLB provision in previous firm as 'very/important' factor in decision to leave previous firm (%)	WLB PULL Good WLB provision as 'very/important' factor in decision to move to current firm (%)
% workers valid response	100	100	85	93	87
DUBLIN + CAMBRIDGE					
All workers (n = 123)	81	3.5	41	33	65
Women only (n = 115)	81	3.6	39	30	65
Working mothers only (n = 45)	84	3.8	36	39	76
UK outside of Cambridge (n = 39)	(74)	(3.3)	(48)	(36)	(68)

Cambridge, average employment tenures are 3.5 years, with 81% of workers classified as mobile (no longer working in their first IT company). Of this mobile worker cohort, one third of workers identified *poor* WLB provision by their previous employer as a 'very important' or 'important' reason for leaving that company. Additionally, two thirds of workers identified *better* WLB provision as a 'very important' or 'important' reason for moving to their current company.[8] Importantly, around half of independent contractors (n = 15) and workers in temporary contracts (n = 16) also identified WLB as an important consideration in moving to their current position. In other words, work-life balance concerns are not the exclusive preserve of workers with permanent contracts. Indeed, the interviews also revealed how some workers had quit to become independent contractors in part driven by dissatisfaction with long work hours and work-life conflict in their previous companies.

Extending the analysis, Table 7.4 shows that whilst WLB provision is an important consideration for the majority of female IT workers surveyed in their decision to move to their current employer, these considerations are most pertinent for working mothers – for example, with over three quarters identifying *better* WLB provision as a 'very important' or 'important' reason for moving to their current company. This is also consistent with some companies' own internal surveys. In Dublin this included one IT multinational's biennial women's survey: 'We asked, "What's the main reason you're leaving the company?" and "What's the main reason you'd stay with the company?" It's work-life balance. And it's increasing … since our last women's survey two years ago, work-life balance went from 28% to 40% as the primary reason why people would stay or leave.' Likewise another Dublin-based MNC in which:

'We'd spent a lot of time and money as a company in developing people, and we were seeing female workers have kids and then leave … There was all this talent and knowledge of all the processes we'd put in, all vanishing out the door. So we developed a working from home policy. Did it impact our turnover? Absolutely … we put in homeworking and turnover came down by 25%.' Director of Human Resources, male, two children, IT MNC, Dublin

Extending these observations to broader bundles of WLB provision, the IT employer survey (150 companies, employing over 8,000 workers locally in Dublin/ Cambridge) measured managerial perceptions of the impacts of total work-life provision and take-up on different dimensions of labour turnover for the period 2004–2007. Here, 63% of managers indicated improved company image to potential recruits; 52% increased retention of women post-maternity leave; 44% increased workforce diversity; and 36% increased female recruitment. Importantly, these figures are higher for firms with more comprehensive suites of WLB provision.[9] These managerial perceptions of positive change are also consistent with measured increases in these same firms' total female workforces over the same time period, and with documented reductions in labour turnover (see James 2014). Indeed, research by Working Families has also documented similar recruitment outcomes as a function of enhanced work-life provision amongst firms beyond the high-technology sector (Swan *et al.* 2011).

Work-life balance and the quality of mobile knowledge, skills and expertise?

In combination, the interview and survey data show how work-life balance considerations actively shape IT workers' decision-making processes around cross-firm mobility, and that firms' provision of work-life arrangements offers a means for them to tap into cross-firm knowledge spillovers and communities of practice. Critics might argue, however, that the implications of this for technical knowledge spillovers are potentially limited given that many women in IT tend to be horizontally segregated in 'female-appropriate' human resources, marketing and other support roles, rather than in 'appropriately-male' technical roles (Ghoshal and Passerini 2006). In response, Table 7.5 shows that in fact over three quarters of the female IT worker cohort in Dublin and Cambridge were employed in technical managerial, technical, or research roles at the time of the survey. They also evidence impressive accumulated employment experience across multiple employers in the technology sector – this as a function of the purposive sampling strategy employed in the IT worker survey, and its administration via the Girly Geekdom and Women In Technology listservs. Thus, of the mobile workers who identified poor WLB provision in their previous company or better WLB provision in their current company as a 'very important' or 'important' reason for moving (Table 7.4), each brought to their new employer an impressive average 11 years' accumulated experience in the IT sector, and of doing IT work in an average of four different companies previously. And while half of these workers had a Masters degree or PhD, this figure increases to over 70% when considering working mothers in isolation.

In short, gendered identities, responsibilities of care and personal life interests beyond the workplace are crucial for understanding the cross-firm mobility of *high-quality* embodied technical and managerial knowledge and expertise in high-tech regional economies. Indeed, IT employers in Dublin and Cambridge also recognised the ways in which WLB provision enabled them to access new sources of external knowledge beyond their existing internal competencies. As neatly summed up by one male CEO in Cambridge, 'it's very high on the agenda because life comes before work for an awful lot of people. And certainly parents who are caring for children, there is no question of them taking employment if it does not present them with the ability to do what actually is their first priority. The flexibility plainly affords us a greater talent pool'. These processes of WLB-informed cross-firm mobility and embodied knowledge transfer therefore help to

Table ... ...reting the quality of their...embedded knowledge in Dublin and Cambridge (n = 115).

ROLE	Current job titles	n	%	IT employment experience (mean yrs)	Total IT employers to date (mean)
SENIOR MANAGEMENT	Chief Executive Officer, Manager Contributions EMEA, Director, Senior Manager, International Service Manager, Chief of Staff, Head of Technology, Executive Director, Senior Programme Manager, Head of Department, Director Product Management, Managing Director	16	14	13.5	4.3
TECHNICAL MANAGERS	User Management, User Experience Manager, Senior Trainer, Service Manager, Senior Project Manager, Project Manager, Assistant Director of Product Development, E-learning Services Manager, Implementation Services Manager, Product Manager, Special Projects Manager, GIS Project Leader, Product Line Director, Senior Technology Project Manager, Program Manager, Service Desk Manager, Customer and Supplier Service Manager, QA Manager, Senior Project Manager, Transformation Programme Manager, Deputy IT Director	33	29	11.8	3.3
TECHNICAL	Software Engineer, MSI Packager, Software Tester, Learning Technologist, Java Developer, Software Architect, Test Team Leader, Software Engineer, Applications Support Specialist, Software Engineer, IT Strategist, Consultant, QA Tester, Consultant, Mechanical Engineer, Quality Systems Coordinator, System Administrator, E-learning consultant, Freelance HTML/CSS Designer, Test Consultant, User Interface Developer, IT Professional, Network Engineer, Senior Consultant, Systems Administrator, IT Consultant, MIS Analyst/Developer, Software Developer, Method Consultant, Web Developer, Web Manager, Web Producer, Web Officer	43	37	9.6	3.6
RESEARCH	Business Analyst, Research Associate, Lead Business Analyst, Business Analyst, Researcher, Business Analyst, Research Fellow, Business Analyst	12	10	13.3	2.9
HR	HR Systems Administrator, HR Professional	2	2	15.0	3.5
MARKETING	Marketing Manager, Marketing, Product Marketing Manager, Marketing Manager, Business Communications Executive	5	4	11.2	3.8
OTHER	Associate, Counsel, Owner, Lecturer	4	3	14.8	2.0

Male survey cohort: Director, Solutions Architect, Software Engineer, Software Engineer, Web Developer, Software Engineer, Technical Support explain why 54% of companies responding to the employer survey indicated 'an improved corporate environment for learning and creativity' as a function of their WLB provision. Significantly, these managerial perceptions were also consistent with measured improvements in firm performance for these same companies over the same time period (2004–2007) (see Chapter 6). In short, WLB provision also has *cross-firm* learning benefits, alongside the intra-firm mechanisms identified in Chapter 5.

Reconsidering Cross-Firm Embodied Knowledge Spillovers

While economic geographers 'have generally written in a normatively favourable if implicit way about the virtues of NIDs [clusters] as providers of good jobs and long-term stability and dynamism' (Markusen 1996: 296), the alternative analysis developed in this chapter reveals a darker side to regional learning. Alongside the cross-firm learning benefits that can accrue to firms through embodied knowledge spillovers, it also shows how patterns of cross-firm job-to-job labour mobility are also often premised on a gendered *dis*satisfaction with work-life conflict, unequal divisions of household labour, uneven and often inadequate WLB provision by employers, and worker concerns around care and a better quality of life. Indeed for some IT workers, work-life concerns also *constrain* inter-firm job-to-job mobility as they seek to avoid the potentially disruptive impacts of a new job on delicately balanced commuting patterns combining nursery, school and a partner's commute; redistribution of domestic duties; possible relocation of home and school; and disruptions to established networks of family and community support. In so doing, this labour geographies analysis disrupts the widely espoused, firm-centric, regional learning mantra that 'labour mobility is always and everywhere a good thing'.

Importantly, these gendered patterns of cross-firm worker (im)mobility and of the role of WLB provision in making firms more attractive to highly skilled workers were identified in *both* Dublin and Cambridge. However, it is also important to recognise some subtle geographical differences in their mode of operation, this as a function of different urban geographies of (dis)connection between sites of production and reproduction. It is also important to locate these two regions within the national economies of which they are part, to overcome an earlier tendency in the regional learning literature to underemphasise the importance of wider extra-local structures in shaping geographies of knowledge spillovers (MacKinnon et al. 2002; Markusen 1999), based on a misplaced conception of regions as 'closed systems' or mere 'containers of intangible assets and structures' (Yeung 2005: 47).

As explored in Chapter 5, IT workers in Dublin generally exhibit greater levels of work-life conflict than do their colleagues in Cambridge. For example, while half of the female IT workers surveyed in Dublin were unsatisfied with their current WLB, this compares with one third in Cambridge. And while 8% of female IT workers in Dublin indicated that they are 'very satisfied' with their current WLB, this figure is twice as high in Cambridge. These patterns are also consistent with a series of specific negative work-to-home spillovers (inability to relax and forget about work; difficulties fulfilling family responsibilities because of time spent working; missing out on leisure time or hobbies; difficulties finding time to maintain friendships), all of which were experienced by a higher proportion of female IT workers in Dublin than in Cambridge on a daily or weekly basis.

Underpinning these differences, research participants highlighted the important role of urban sprawl associated with strong house price increases over the last two decades, which have forced many workers in Dublin to move out of the city in search of more affordable housing, yielding commutes of up to 4 hours per day in some cases. In contrast, such lengthy commutes were rarely identified at interview among the Cambridge research participants. The IT worker survey also documented similar disparities, with almost three times as many Dublin workers surveyed commuting 3 or more hours per day (or 15 hours plus per week) than their Cambridge colleagues. Against this backdrop, employer provision of working-from-home arrangements (as a means for reducing lengthy commuting times) was typically identified as having a greater meaning and significance for research participants in Dublin than in Cambridge, with significant implications for workers' patterns of decision-making around cross-firm job-to-job mobility.

Reinforcing these differences, Ireland and the UK also evidence different gendered welfare regimes in relation to paid work and care.[10] There is no statutory provision for paternity leave in Ireland and no legal right to part-time work, job sharing, flextime and teleworking (all at employers' discretion) (Russell et al. 2009). In addition, the statutory paid maternity leave entitlement is lower in Ireland than in the UK, and Ireland exhibits higher costs of childcare in relation to average incomes. Against this backdrop, employer-provided extra-statutory maternity leave, paternity leave and flexible working arrangements also have an enhanced meaning and significance in Dublin than in Cambridge, with consequent effects for patterns of cross-firm job-to-job mobility. Thus while IT employers in Dublin reported 'increased female retention post-maternity leave' (57%) and 'increased workforce diversity' (49%) as a function of their total bundles of WLB provision (2004–2007), this is less apparent amongst the Cambridge IT companies surveyed (with equivalent figures of 42% and 37%).

The analysis also identifies the significance of 'luxury' concerns around work-life balance – alongside the recessionary rise of non-standard employment contracts, survivor shock, weakened employment protection, layoffs and non-self-motivated firm exit – in *continuing* to shape patterns of job-to-job mobility by female technology workers. Within the sample of female IT workers surveyed in December 2010 (n = 139),[11] over half (78) had changed employers in the previous three years. Of this mobile worker cohort, 31% of workers (cf. 37% of mothers with dependent children) identified *poor* WLB provision by their previous employer as a 'very important' or 'important' reason for leaving that company.

Additionally, 42% of workers (cf. 58% of mothers with dependent children) identified *better* WLB provision as a 'very important' or 'important' reason for moving to their current company. Only 17 workers identified WLB provision as 'irrelevant' in their decision to move to their current IT employer in the recessionary context. The results also highlight the role of layoffs in shaping patterns of embodied knowledge spillovers through cross-firm worker mobility in a manner under-researched in the capital-centric regional learning literature. As testament to the significance of this dark side of regional knowledge spillovers, one third of the workers surveyed had experienced a period of involuntary unemployment in the previous three years.

Overall, then, we need to recognise that the mobility of embodied tacit knowledge, accumulated skills and expertise between firms within high-tech regional economies does not operate on some homogeneous, flat surface. Nor can we continue to position firms as the only relevant institutions, between which workers move in a straightforward manner driven only by technological innovation concerns. Rather, cross-firm knowledge spillovers are unavoidably shaped by employer-provided WLB arrangements, uneven gender divisions of household care and delicately balanced work-care trajectories through the city incorporating public transport networks, school runs, crèches, partners' workplaces and a myriad of other infrastructures of everyday work-life. Once again, these everyday work-life infrastructures represent more than just a benign social reproductive 'stage' upon which cross-firm knowledge spillovers are played out, but are fundamental to how those spillovers are *constituted* (and indeed not constituted) by workers in the first place. It is therefore imperative that economic geographers bring gendered geographies of work-life conflict, social reproduction and urban carescapes to the *core* of an expanded regional learning and innovation agenda, through new intellectual engagements with a rich body of work in feminist geography.

Notes

1   Indeed, for Richard Florida 'such hypermobility has become the norm across the entire economy' (2002: 104). Arlie Russell Hochschild (1997: xix) offers a useful commentary on the origins of the 'zero drag' moniker amongst Silicon Valley engineers in the late 1990s.
2   While Sweden has one of the highest female employment rates in the OECD, parenthood pushes a higher proportion of mothers (18%) than men (10%) to seek part-time employment (OECD 2012).
3   Hanson and Pratt conducted in-depth interviews with men and women in over 600 households in the Worcester Metropolitan Statistical Area to generate an area-based sample of workers that was representative of the Worcester working-age population.
4   Thus for Hanson and Pratt (1991), 'women end up in female-dominated occupations not because they are making rational, long-term, life-long income-maximizing decisions, but because they are faced with severe day-to-day space-time constraints, dictated in part by their domestic workload' (p. 251).

5   Similarly, Mitchell et al. (2001) set out how 'Organizations face many costs directly related to turnover, including exit interview time and administrative requirements, payout of unused vacation time, and the cost of temporary workers or overtime for co-workers asked to fill in. Replacement costs include advertising, processing of candidates, interviewing, and selection. Finally, training costs – both formal and informal – add to the overall burden' (p. 96).

6   WLB provision scores were calculated simply for each firm by adding up the different types of WLB arrangements available to employees at the time of the survey (see Table 7.4). Scores ranged from a maximum of 16 down to 0, with a mean average of 4.

7   Encouragingly, participants also pointed to the potential for positive change over time: 'The next generation of managers coming through, they're more comfortable with bringing in the work-life balance provisions. Because they see it gets the best out of their people. Partly because they have gone and tried it for themselves.' Principal IT Consultant, female, IT SME (previously at US-owned IT MNC), UK SE region.

8   65% of IT workers surveyed in Dublin and Cambridge identified the significance of WLB provision in shaping their decision to move to their current employer. Compare this with an equivalent figure of 41% in the UK's Fourth National Work-Life Balance Survey (2011), based on a much larger sample of 2,767 workers across multiple sectors.

9   Cambridge and Dublin IT employers combined (excluding firms with WLB provision scores less than 4 (mean average)) (n = 86): 72% indicated 'improved company image to potential recruits'; 63% 'increased retention of women post-maternity leave'; 45% 'increased female recruitment'; and 44% 'increased workforce diversity'.

10   Previous research has made clear that the terms in which the WLB agenda is cast vary between nation-states as a function of different national production regimes, labour market regimes, working-time regimes and welfare state regimes (Crompton 2006: 116; Gambles et al. 2006; Lewis 2009).

11   While the majority are UK based, these workers exhibit a similar skills and competencies profile to the 2008 IT worker survey sample with an average 11.6 years' work experience in the IT sector across 3.8 different employers. Additionally, 39% had a Masters degree or higher; 70% were employed in technical managerial, technical, or research roles at the time of the survey; and only 7% in HR, marketing or sales roles. The 2008 and 2010 survey samples also exhibit similar age distributions and proportions of working mothers with dependent children of pre-school age and in full-time primary education.

Chapter Eight
Conclusions: Gendered Regional Learning and Work-Life Advantage

Introduction

Over the last three decades, economic geographers have analysed in ever finer detail how the spatial co-location of high-tech firms in regional industrial agglomerations helps foster learning, innovation and economic competitiveness. Without doubt, this near obsession in economic geography has yielded important insights into the socio-cultural foundations of regional advantage, informed regional economic development policies worldwide, and inspired multiple generations of economic geographers. Yet despite its concerns to 'demystify' (Gertler 1997) the cultural constitution of economic activity, this influential body of work remains strangely gender-blind. Beyond a few maverick studies, the regional learning monologue has paid almost no attention to how gendered identities, varied responsibilities of care and personal life interests beyond the workplace shape workers' (non-)participation in the relational networks and communities of practice widely theorised as enabling regional learning and innovation. Rather, these factors are ignored, as best left to feminist scholars, who for their own part are often dismissive of the regional learning research agenda as myopic in its firm-centric, masculinist and economistic focus. Barnes and Sheppard (2010) have characterised this situation as one of 'fragmented pluralism' and a lack of productive intellectual conversation between different sub-disciplinary strands of human geography. The overall result is that the wider consequences of economic change and firm competitiveness for the well-being of people in regional

Work-Life Advantage: Sustaining Regional Learning and Innovation, First Edition. Al James.
© 2018 John Wiley & Sons Ltd. Published 2018 by John Wiley & Sons Ltd.

economies are correspondingly neglected (Perrons 2001), in a manner that distances economic geography from growing international debates around socially inclusive growth, labour market inequality and worker well-being; and hence questions the wider social relevance of the regional learning agenda itself.

This study has, for the first time, brought the expansive regional learning research agenda into conversation with the equally expansive work-life balance (WLB) research agenda, which is concerned with the shifting temporal and spatial boundaries between 'work' and 'home', and how to enable workers and families to reconcile paid work and social reproduction in pursuit of improved well-being. Importantly, both research agendas ultimately respond to the emergence of post-Fordist flexible production processes from the late 1970s, one exploring the regional territorial forms of flexible production, and the other dramatic changes in the organisation of 'flexible' paid work and working times. *Work-Life Advantage* has explored the unavoidably gendered everyday geographies of work-life, learning and innovation amongst information technology workers and firms in Dublin, Ireland and Cambridge, UK prior to, and subsequent to, the onset of the Great Recession in 2008. Debates around work-life balance have come to assume a strong national significance in both countries over the last two decades, and Dublin and Cambridge are both widely recognised as important European 'blueprint' clusters of high-tech growth of interest to policy-makers elsewhere.

Within this research framework, the ambition of the analysis developed within *Work-Life Advantage* was fivefold. First, to make visible the gendered networks of social reproduction and household divisions of labour in which female and male knowledge workers are intimately embedded, and which unavoidably shape their abilities and willingness to perform as high-quality 'human capital' factor inputs to firms' knowledge production and exchange processes on a daily basis. Second, to explore how far the everyday workplace practices which collectively underpin the learning and innovation mechanisms widely celebrated in the regional learning literature might also be socially damaging, with negative implications for workers' health, well-being and quality of life across the work–home boundary. Third, to identify different types of working arrangements and employer interventions which might meaningfully reduce those challenges for workers and their families, and in a manner that also enhances firms' learning and innovative capacities and long-term sustainable competitive advantage. Fourth, to explore the role of uneven levels of work-life provision by different high-tech employers in motivating and constraining patterns of cross-firm job-to-job mobility, and hence spillovers of workers' embodied knowledge and skills. And fifth, to explore the spatial variability of high-tech work-lives, within and between different regional economies and national welfare regimes.

The first half of this concluding chapter summarises the book's major substantive findings, core arguments and intellectual contribution in relation to these aims. The second half identifies future research possibilities, which include shifting the dominant focus of the WLB literature concerned with how best to flexibilise work around an assumed female majority responsibility for childcare, also to explore how best to support men who are 'going against the grain' (Ranson 2010); and extending the analysis to workers and firms in the Global South, in order that we begin to provincialise the 'universal' work-life balance concept which originated in the Western context (Lewis et al. 2007).

Taking Back Work(-Life)?

There is no sense that the work-life balance problem has been 'solved'. (Gatrell and Cooper 2008: 72)

There's not enough time for life. We are working more but surviving poorly. We need to reconsider our working lives in the context of our own well-being and the well-being of other humans. (Gibson-Graham et al. 2013: 18–19)

As I introduced in the first part of this book, the ascendancy of the work-life agenda over the last three decades emerged from growing concerns around 'the increasing pervasiveness of paid work in people's lives' (Lewis et al. 2003), and the complex, gendered and often painful set of conflicts between competing demands of paid work and personal responsibilities and life interests beyond the workplace (including care for dependants, friendships, communities, personal life interests and leisure) (Bunting 2005; Weeks 2011). As outlined in Chapter 3, these work-life challenges can be understood as a function of 'work intensification' (Burchell et al. 2002), reinforced by weakened employment laws, an erosion of trade union power, tyrannies of performance management and employee perceptions of job insecurity that further motivate them to work harder. At the same time, household life has also become more complex through increased female labourforce participation and increased numbers of dual-earner households. And on top of all this, the neoliberal attack on social provisioning has transferred the burden of care down to the 'natural' level of home (Bakker and Gill 2003) where most women continue to bear the prime responsibility for the 'messy and fleshy' components of domestic and family life.

In response, the work-life research agenda considers how to redistribute paid work and care through 'non-traditional' working arrangements that change the conditions under which both women and men work; to challenge long-standing masculinist definitions of 'the ideal worker' rooted in presenteeism; to humanise the workplace for all workers who wish to lead rounded lives; and to challenge a growing 'care deficit', based on a shortage of time and energy to invest in nurturing resilient, secure individuals, families, friendships and communities. In this way, the work-life agenda overlaps strongly with Gibson-Graham et al.'s (2013) wider geographical agenda to *Take Back the Economy*. Notably, in Chapter 2, 'Take Back Work – Surviving Well', Gibson-Graham et al. offer a critique of the ways in which paid work is taking over some people's lives, and the need to limit the time we spend in paid work and fashion different work futures so that there is time left for a dignified life. They argue that we need to 'take back work' as a means to an end that enables us to engage in other meaningful and fulfilling parts of life (recreation, hobbies, family, friends, community, charity, study, sports), rather than paid work becoming an end in and of itself that crowds out other areas of life.

However, in common with much of the work-life literature, Gibson-Graham et al.'s argument is quite compelling on *why* we need change as a function of work intensification and paid work crowding out other meaningful areas of life, but less so in terms of outlining *how* to promote more socially sustainable forms of working in practice. In the absence of widespread radical downshifting of the kind advocated by Gibson-Graham (and its associated reduction in income for workers and their families), key questions remain about *how* to encourage employers to implement 'alternative' working arrangements that better enable workers to reconcile competing activities of work, home and family. As explored in Chapter 3, while multiple studies have convincingly demonstrated the causal connections between poor work-life balance and increased stress, reduced psychological well-being, deteriorating familial relationships and ongoing gendered labour market inequality, work-life provision continues to be seen by many employers as too costly and disruptive (especially for small firms) and/or unfairly privileging a small subset of their workforce. Accordingly, any workers who *do* make use of WLB arrangements are often perceived negatively as less committed and adversely affecting the firm, and requests for alternative working arrangements are often turned down on the basis that they will harm firms' operational effectiveness.

The so-called 'business case' for WLB lies at the heart of government policy interventions in this area in liberal welfare states, and is epitomised in the UK, Ireland and the USA. Yet, as I argued in Chapter 3, there remains a relative dearth of empirical evidence to support government claims around the economic benefits of WLB provision in practice. The impact of existing WLB studies has also been limited by (i) a prioritisation of employer needs at the expense of workers and their families; (ii) a limited focus on *output* measures of firm performance rather than on firms' underlying capacities for learning and innovation; and (iii) atomisation of firms from the regional industrial systems of which they form part. These three limitations frame the alternative regional learning/ WLB mutual gains analysis developed in Chapters 5, 6 and 7 concerned to bring regional learning, labour geography and feminist economic geography into new productive conversation. The four major contributions of this hybrid analysis are outlined below, as part of a larger agenda that builds deeper connections amongst human geographers through engaged pluralism.

Making Space for Female Worker Agency

After three decades of development and refinement, the regional learning and innovation literature now offers a wide variety of territorial innovation models, as successive cohorts of scholars have differently sought to conceptualise the learning advantages of 'being there'. Strikingly, however, the majority of analytical insights to date have emerged from empirical studies of male knowledge workers, or else from analyses that collapse all workers into unitary, genderless categories of 'labour', 'human capital' or 'competence portfolios'. Previous work has said almost nothing about the lived experiences and active agency of female knowledge workers in (re)producing learning and innovation activities in regional industrial systems, or of the gendered social relations in which workers and firms are embedded. As I argued in Chapter 2, these exclusions and silences might in part be understood as rooted in the male-dominated sectors on which regional learning scholars have focused their attention (including computer software, semiconductors, motor sport industry, engineering and biotechnology) reinforced by stubborn social constructions of high-tech industries as *masculine*. And while a select number of studies *have* included gender as a relevant variable in their logistic regression analyses of regional knowledge spillovers through cross-firm worker mobility (Dahl 2002; Thulin 2009), this has typically been in the form of a female 'dummy variable' that reduces female technology workers to that which they are not ('minus male').

In response, the analysis developed in the second half of this book has been concerned to expose the masculinist myopia of the regional learning and innovation agenda and attendant theories of regional advantage. In Ireland and the UK (two empirical heartlands of the regional learning agenda) women account for almost one third of the high-tech workforce (EU Labour Force Survey 2014). The analysis pointed to the crucial role of women in staffing and managing hundreds of different IT firms in the UK and Ireland – this as a function of increased female labourforce participation in STEM occupations, increased female educational achievement, and shifting societal and welfare constructions of the 'good mother' as she who enters the labour market to raise her income and skill levels for the benefit of her children, who no longer occupies the home as a continuous presence, and who hands over the care of her children to another for part of the day (Perrons et al. 2006). And beyond their labour market penetration into 'female-appropriate' HR, marketing and other support roles (see, e.g., Ghoshal and Passerini 2006), the analysis evidenced the presence of over three quarters of these women in technical and technical-managerial positions.

On one level, these identified patterns of female worker agency challenge the false regional learning dualism between a male-occupied rational sphere of transcendence (science makes breakthroughs, is involved in change, in progress) and a female-occupied emotional sphere of immanence (the static realm of living in the present, or simple reproduction). And much more than 'human capital' as an abstract input to firms' knowledge production processes, IT workers are actual human beings embedded in multiple gendering and sexing practices, which define the terms upon which men and women supply their labour (and also motivate workers actively to move between firms in pursuit of an improved work-life balance). As shown in Chapter 5, these practices include the asymmetrical 'identity risks' which emerge from pursuing parenthood alongside professional knowledge roles and place particular pressures on women: the identity of being 'a good mother' invokes an everyday presence and involvement in childrearing that is absent from dominant societal expectations of what constitutes 'a good father' (Hardhill and van Loon 2006). In the absence of appropriate WLB arrangements, Chapters 5 and 7 showed that some women take compromise jobs which reduce work strain but ultimately under-utilise their skill-sets, knowledge and experience on behalf of their employers (constrained female worker agency). Indeed, some women subsequently quit IT employment altogether, resulting in a substantial loss of embodied female skills, training investment and accumulated experience in ways far less common among their male colleagues – yet whose experiences continue to define successive advances in regional learning theory.

In these ways, this study gives voice to women as a crucial set of agents who have largely been rendered invisible in the regional learning literature to date, and demands that other scholars follow suit. The aim, then, is to extend earlier feminist critiques of 'universal' economic theory to the heterodox regional learning agenda, and thereby to 'bring it down to earth and give [it] a pair of pants' (Bordo 1990: 137). In short, it is simply impossible to explain the uneven geographies of regional innovation and learning unless we take into account the labour agency of female technology professionals, who do more than simply watch passively from the sidelines while their male colleagues do all the hard work. Much remains to be done to delineate the learning and innovation (dis)advantages of 'being there' through the eyes of female innovators.

Connecting Regional Worlds of Production to Regional Worlds of Social Reproduction

Closely linked to the general analytical silence around gender relations and female worker agency within regional learning research is a widespread failure to locate activities of knowledge production within the wider networks of social provisioning and care that enable *socially sustainable* innovation and worker learning in practice. A core premise of the arguments developed throughout the book is that we cannot conceive of knowledge workers as acting autonomously; as if the learning activities and workplace interactions in which they engage on a daily basis 'can be separated analytically and ontologically from their wider existence' (Castree 2007: 859). Extending earlier critiques of regional economies as 'closed systems' or mere 'containers of intangible assets and structures', the analysis has made clear workers' fundamental inseparability from a myriad of hidden household caring activities, in which they are simultaneously embedded alongside their everyday participation in learning and innovation activities in the formal economy. The analysis has also documented the everyday tensions (work-life conflicts) that emerge from that inseparability, with which many workers struggle to juggle in the absence of meaningful support.

Divergent from influential earlier accounts of young, unattached Silicon Cowboys engaged in long hours working driven by 'the sheer thrill of work' and 'a love of technology' unhindered by any demands of family or personal life, the analysis has identified a more complex set of worker experiences. In Chapter 5, we heard the testimonies of male engineers with significant childcare responsibilities who are no longer able or willing to buy in to these heroic types of high-tech work culture as they seek to reconcile shifting societal expectations around active fatherhood with peer pressure from single male co-workers and managers without children. Likewise the too-often unheard voices of female technologists, for whom the temporally variable demands of IT work across product development and sales cycles generate significant work-life conflict as they try to juggle those demands with personal life responsibilities and commitments whose timings are instead fixed and regular. These include daily school runs, daily picking up partner from work, weekly Cub Scout leadership, weekly amateur team sports fixtures, parents' evenings, assemblies, homework and evening classes. Tensions also arise in relation to the fixed schedules of school holidays and term times. Problematically, the pressure to demonstrate commitment by working long hours is also likely to be strongest in the early stages of the career, at the same time as many workers are starting families and are concerned not to miss out on children growing up.

And as an outcome of these tensions, Chapters 5 and 6 suggested that the very working arrangements and workplace practices that are widely heralded to underpin the competitiveness of knowledge-intensive firms through time-based competition are in fact socially damaging and unsustainable. The most commonly cited negative work-to-home spillovers include persistent feelings of guilt at missing out on family meal times, bath times and bedtime stories, interrupted sleep patterns, stress and exhaustion impacting on relationships with children and partners, working when feeling unwell, missing out on leisure time and hobbies, and an overall reduced quality of life. Care deficits were also manifest through self-identified patterns of crossover stress in dual-earner households, as partners inevitably pick up the care tab that people have not been able to pick up themselves through over-commitments at work.

These work-life conflicts were also identified in the second half of the book as motivating many tech workers to seek flexible working arrangements to effect change for the better. As outlined, employer-provided work-life arrangements are typically grouped in terms of their providing: (i) greater flexibility of *when* work is done (e.g. flextime, annualised hours, compressed work weeks); (ii) greater flexibility in the spatial location of *where* work is done (e.g. working from home); (iii) a reduction in total work hours (e.g. 80% FTE contracts); and (iv) workplace social support for parents, including employer assistance with childcare. The analysis showed that there is no single WLB panacea; rather, technology workers' work-life preferences vary significantly within gender groups by job function, department and household situation. Consequently, different home and work factors mean work-life provisions suitable for one class of employees may have little or no effect in reducing work-life conflict for another class. Further reinforcing this complexity, the kinds of WLB provisions required by a particular individual at one point in the lifecourse are also changeable over time in relation to major life events, including labour market entry, marriage, divorce, childrearing, moving house, nest-leaving and eldercare responsibilities. Ultimately, this diversity points to the urgent need for more comprehensive suites of employer-provided work-life arrangements.

Chapter 6 also identified three sets of positive outcomes amongst IT workers making use of employer-provided work-life arrangements in practice (part I of a mutual gains research agenda), focused on IT firms with more comprehensive patterns of WLB provision. First, *reductions in the intensity and frequency of time-based conflict*, in which time spent by individuals doing paid work reduces time available for carrying out other family responsibilities and personal life activities. Second, *reductions in strain-based conflict*, in which home and family interactions are negatively affected by stress and anxiety resulting from activities in the paid work domain. And third, *reduced cross-over stress* in which the work-life conflict experienced by one spouse impacts on the other spouse, as the latter increases their own unpaid work in response to the paid work demands placed on the former. However, as also explored in Chapters 5 and 6, there are important differences between work-life arrangements that accept the accommodation of working time to the gendered division of labour and those that also challenge the male carer model at a deeper level by supporting the reallocation of reproductive tasks within the household. Employer preferences tend to remain, for now at least, with provision of the former.

Work-Life Balance as Regional Learning Advantage?

Despite the profound moral and social significance of work-life conflict as a threat to the well-being of workers, families and communities, employers have been widely identified as unlikely to implement meaningful work-life support arrangements unless they can identify bottom-line economic advantages that arise from doing so. These sentiments were also echoed by managers in Dublin and Cambridge (see Chapter 6). Previous studies of the organisational benefits of work-life provision have tended to restrict their focus to quantitative output measures of firm performance (especially output per employee, labour turnover and absenteeism), which of themselves say little about firms' capacities for sustainable competitive performance. As I have argued throughout this book, firms' innovative capacities are not discrete, abstract, disembodied processes. Rather, they are *peopled*, by real workers with social identities, family commitments and extra-curricular interests, who interact on an everyday basis to combine varied skills, competencies, ideas and prior experience to create new knowledge and apply it incrementally in the pursuit of improved economic performance. As such, they are *unavoidably* shaped by the workplace institutional environment of which work-life arrangements form an increasingly important component. Nevertheless, this is an area in which conventional WLB business case analyses have remained silent.

Chapter 6 explored how, by making available the kinds of WLB arrangements identified by technology workers in Chapter 5 as offering meaningful reductions in gendered work-life conflicts, there can also result beneficial impacts on firms' innovative capacities (part II of a mutual gains research agenda). At the broadest level, over half of the managers responding to the employer survey (N = 150) indicated 'an improved corporate environment for learning and creativity' as a result of their total bundles of WLB provision – perceptions that were also consistent with measured improvements in firm performance over the same time period. Only three firms (2%) surveyed identified a negative impact of WLB provision on their 'corporate environments for learning and creativity'.

Underpinning these WLB learning advantages, three concrete mechanisms were identified in Chapter 6. First, participants emphasised the learning advantages of WLB arrangements which give them *increased autonomy, freedom and independence* over the spatial location and/or temporal pattern/sequencing of their work hours, in a manner that better enables them to engage in individual cognitive activities and to use their time more effectively and in less fragmented blocks, at times which often do not coincide with 'normal' office hours. Second, participants highlighted how comprehensive bundles of work-life provision can enhance work team diversity by means of retaining highly qualified IT professionals (especially female returners) with significant caring responsibilities and commitments outside work, some of whom had previously considered leaving due to recurrent problems of time-based and strain-based work-life conflict. In this way, WLB provision can *widen firms' repertoires of perspectives and skills*, *networks of external contacts* and hence capacities for comprehensive problem solving in the face of new events. Third, employer-provided WLB arrangements can foster an *increased longer-term sustainability of worker learning*. Here research participants described the ways in which time pressure and long hours working can also result in self-identified reductions in their abilities to work effectively and to engage in the everyday activities of problem solving, coordination and symbolic analysis which define their work as IT professionals. Participants described how, by repeatedly imposing too great a workload in too short a timeframe, the motivational benefits of time pressure in the short term give way to stress and fatigue, which undermine creativity and the quality of decision-making. The results from the IT worker survey (N = 162) are also striking, with 94% of workers indicating that, as a result of making use of their personally preferred employer-provided WLB arrangements, they were less stressed at work. In addition, 79% reported greater engagement with their work; 78% that they could think more creatively at work; and 75% that they could learn more effectively at work. These figures are even higher for working parents.

In addition to these WLB learning benefits a number of challenges also arise. One set of challenges concerns the *differential ability of workers to act as 'independent learning agents'* and to have the confidence to take informed risks and make errors in ways widely identified in the learning literature as enabling innovative capacity. Significantly, Chapter 6 pointed to a clear perception amongst several workers engaged in homeworking, flextime and reduced hours that their behaviour is being *more* closely monitored than colleagues whose commitment to the company is understood as unproblematic given their constant physical presence at work during 'normal' work hours. A second set of challenges concerns the negative implications of homeworking, reduced work weeks and flextime for undermining the frequency and quality of face-to-face communication between co-workers, undermining transmission of crucial non-verbal cues which otherwise allow team members to adjust their individual and collective behaviours to reduce ambiguity and uncertainty arising from complex tasks. Importantly, the analysis also suggests that these negative outcomes are not inevitable. Rather, they are differently mediated by managerial buy-in and work group ratification of non-standard working arrangements and the use of information communication technologies (e.g. online team meetings and brainstorming sessions using instant messaging and video conferencing facilities) to maintain work group capacities for multi-directional, simultaneous cycles of discussion, interruption, repair, feedback and learning quickly *at a distance*. Thus, rather than understand the WLB business case only in terms of any net positive impacts on the firm, there is a need to broaden the definition to include a *lack of any negative* impact.

The wider significance of this analysis then relates to the 'global' economic downturn, which has created new work-life demands. And with many employers keen to effect cost savings in pursuit of post-recessionary survival and growth, employer-provided work-life arrangements have been far from immune (Galinksy and Bond 2009). Likewise the multiple CEOs who took part in this study for whom work-life provision was identified as merely an additional cost to the firm, and variously described as 'prohibitive', 'burdensome', 'a significant challenge', 'causing work-flow problems' and a 'management headache' that 'needs to be reined back in'. These views were most commonly articulated in smaller IT firms with limited resources. A core argument developed in the latter half of this book is the irony of the rollback of WLB provision by employers as short-term cost saving in the wake of recession. Such rollback risks undermining the very mechanisms identified here as *enabling* firms' capacities for learning and innovation as sources of long-term competitive advantage and growth.

Exposing the Gendered Foundations of Regional Knowledge Spillovers

The model of the knowledge production function from the literature on innovation and technological change can be represented as:

$$I_i = \alpha RD_i \beta HK_i \gamma \varepsilon_i$$

where I stands for the degree of innovative activity, RD represents R&D inputs, and HK represents human capital inputs [!]. Audretsch and Feldman (2004: 2716)

A distinctive feature of the mainstream work-life research agenda outlined in Chapter 3 is its predominant analytical focus on single case study firms, too often presented as independent entities and abstracted from the regional economies in which they are located. Consequently, we know very little about the effects of work-life provision on networks of interaction *between* firms. One widely identified mechanism for the cross-firm transfer of knowledge is the physical mobility of highly qualified workers ('learning-by-hiring'). As employees move between workplaces, comparisons of evolving ideas are made with 'how things are done in other firms' (Henry and Pinch 2000: 198); and new constellations of talent increase the potential for learning synergies. And much more than a once-and-for-all knowledge transfer, mobile workers may also maintain advantageous *ongoing* knowledge links between their new and previous employer. Problematically, however, cross-firm job-to-job mobility is overwhelmingly understood in the regional learning literature as an 'easy and rapid movement' (Fallick et al. 2004: 2) carried out by seemingly independent and genderless 'innovation agents' or 'human capital inputs' (represented as HK in the opening quote) who simply switch between firms driven by technological imperatives. Worse still, to reduce these workers merely to 'competence portfolios' (or seemingly disembodied 'mobile brains') means that previous research has said very little about these workers' social identities, about their lived experiences of (re)producing knowledge spillovers in practice, nor about the geographical possibilities for them to negotiate work-life improvements in the process.

Against a backdrop of uneven WLB provision by IT employers, the alternative labour geographies analysis developed in Chapter 7 showed how WLB provision can increase firms' abilities to access cross-firm embodied knowledge spillovers relative to firms with lower levels of provision. It also showed how these patterns of labour mobility so widely celebrated in the regional learning and innovation literature are also premised on common experiences of work-life conflict and hardship amongst female and male technology workers; unequal gender divisions of household labour; uneven and often inadequate work-life provision by employers (formal arrangements *and* informal support practices); and worker concerns that stretch far beyond knowledge production ('a love of technology') to include a love of family, reproduction, care and quality of life. The analysis showed that we need to recognise that the widely documented 'carriers of critical knowledge flows' are also wives and husbands, mothers and fathers, daughters and sons, and members of communities and friendship groups. Accordingly, the everyday *abilities* and *willingness* of technology workers to move between firms is far from a unidimensional phenomenon carried out by atomised units of labour driven only by corporate logics to maximise innovative capacity and quarterly revenues on behalf of firms. Crucially, the analysis also showed how work-life concerns *constrain* cross-firm job-to-job mobility amongst female and male technologists alike (especially those with young children), as those workers seek to avoid the potentially disruptive impacts of a new job on delicately balanced 'crazy quilt' (Balbo 1987) commuting patterns combining scheduled and unscheduled childcare needs, partners' commutes, a potential redistribution of domestic duties, possible relocation of home and school, and disruptions to established networks of family and community support. These findings expose the economistic limits of previous regional learning studies in which 'constraints to mobility are often regarded as market imperfections' (Fischer and Malmberg 2001: 358).

As such, the analysis developed in Chapter 7 exposes a darker side to cross-firm knowledge spillovers in regional economies, as experienced by workers and their families, which has previously been obscured. Much more than simply adding gender in as an additional analytical category ('tucked away in an odd corner and presumed to be of interest only to women' (McDowell 2000: 498)), it begins to expose the masculinist biases of embodied knowledge spillovers and knowledge externalities that regional scholars have come to take for granted. Such concepts are found lacking once we reconceive of innovation agents as real people with different gender identities, families, partners, dependants and lives outside work. The analysis challenges the powerful (yet problematic) economistic claim that cross-firm labour mobility is always and everywhere a good thing, motivated by 'the cause of advancing technology' and 'far greater loyalty to one's craft than to one's company' (Saxenian 1994: 36). Moreover, the analysis also documented the role of work-life conflict and uneven WLB provision by employers in *continuing* to shape patterns of cross-firm job-to-job mobility and knowledge spillovers in the wake of recession.

Uneven Geographies of Work-Life Balance

The everyday attempts of workers to achieve a personal work-life balance involve the delicate coordination of multiple tasks of production and household reproduction, which are distributed across different (and often mismatched) spaces and times. IT workers in Dublin generally exhibit greater levels of work-life conflict in trying to achieve that delicate coordination than do their colleagues in Cambridge, a pattern that is evident across multiple worker cohorts. Underlying these differences, Chapter 5 highlighted the difficulties of coordinating everyday life within different urban carescapes, with Dublin and Cambridge evidencing different degrees of spatial (dis)connection between sites of housing, jobs, schools, transport systems, leisure and commodified forms of caring services (eldercare facilities, childcare services, gyms, hair salons). These different urban carescapes are crucial therefore to understanding the wider context in which work-life calculations and decisions (and their implications for appropriate employer responses) are made – or what Jarvis (2005: 141) has identified as the 'infrastructure of everyday life'. In particular, the analysis identified in Dublin the work-life significance of expanding travel-to-work areas, longer and more intense rush hours and growing congestion associated with school runs as some of the major everyday challenges that face workers, families, employers and service providers in struggling to reconcile the daily complexities of social reproduction with waged work around daily activities that are often not in close proximity to one another. In this way, urban infrastructures of everyday work-life are more than just a benign social reproductive 'stage' upon which regional learning and innovation are played out, they are *constitutive of* regional industrial systems, because they are fundamental to workers' everyday abilities to engage in learning and innovation processes on behalf of their respective employers. Yet these are not factors that have figured prominently in most regional learning accounts.

In seeking to understand these work-life (dis)connections, the analysis developed in the second half of the book also stressed the need to locate workers, firms and regional industrial systems within the national economies (and indeed international production chains) of which they are part. This is important because the terms in which the work-life balance agenda is cast vary between nation-states as a function of different national production regimes, labour market regimes, working-time regimes and welfare regimes (Esping-Anderson 1999; Crompton 2006; Gambles et al. 2006; Lewis 2009). This national focus also overcomes an earlier tendency in regional learning analyses to under-emphasise extra-local structures (MacKinnon et al. 2002), rooted in a misplaced conception of regions as 'closed systems' or mere 'containers of intangible assets and structures' (Yeung 2005: 47). There is currently no statutory paternity leave or legal right to work part-time in Ireland; statutory paid maternity leave entitlement is lower than in the UK; and compared with other European countries, Ireland exhibits higher costs of childcare in relation to average incomes. Against this backdrop, employer-provided extra-statutory maternity leave, paternity leave and flexible working arrangements have an enhanced meaning and significance in Dublin than in Cambridge, with important consequences for workers to engage in everyday learning and innovation activities on behalf of their respective firms (Chapter 6), and also motivating some workers to move between IT employers (Chapter 7).

These uneven geographies of work-life conflict are also evident in the wake of recession. Here the analysis in Chapter 5 evidenced negative changes in female and male IT workers' recessionary experiences of work-life conflict as a function of understaffing, higher workloads, more unpaid evening and weekend overtime, less predictable schedules, pay freezes, increased wage dependency on bonus incentives and heightened fears of job loss – these in turn reinforced by austerity budgetary constraints on welfare provision. And once again, at the time of writing these effects seem to be more pronounced in the Irish context of crisis and austerity (see also Ó Riain 2014; O'Callaghan et al. 2015), where earlier worker preferences for reduced hours working as a means of reducing work-life conflict are subsequently less evident because of their associated reduction in pay. The danger, then, is the risk of creating greater inequalities around work-life provision, given employer suggestions that in the wake of the downturn, calls for altering WLB practices may seem a 'little indulgent' (EHRC 2009: 6); and 'the luxuries of a booming economy that cannot be sustained as we seek to recover from recession … and [an] unnecessary hindrance on business performance' (Leighton and Gregory 2011: 11). Accordingly, the reconfigured WLB business case developed in this book – concerned to identify the complementary role of work-life provision in sustaining regional learning and innovation – becomes even *more* salient. It also contributes empirically and conceptually to the 'holistic development' agenda (Pike et al. 2006) concerned with expanding the narrow analytical focus of regional studies beyond economistic – or 'dessicated' (Morgan 2004) – indicators of competitiveness, growth and productivity to include social inclusion, gender equity, quality of life and family well-being. In particular, it offers an extended response to the core challenge of identifying the *practical* and *organisational* means for promoting more socially inclusive forms of learning, innovation and economic growth in a manner that challenges the labour market marginalisation of female technologists in pursuit of more evenly shared prosperity.

Future Research Possibilities

This book offers just one plausible response to recent calls for economic geographers to develop more critical, heterodox and pluralist analyses of economic life that disrupt conventional economic wisdoms and taken-for-granted understandings; to recognise and problematise false portrayals of 'the economic' as an autonomous sphere; and to explore *how* and *where* economies function, for *whom*, and to *what ends* (see Christophers et al. 2016). In response, the analysis developed here has taken one of the cornerstone research agendas that have come to define economic geography as an intellectually and policy-relevant sub-discipline over the last three decades and sought to open up new perspectives. The resulting hybrid work-life/regional learning agenda challenges the false abstraction of networks of knowledge production and exchange from gendered networks of social reproduction and care. And hence it challenges the long-standing analytical silences by economic geographers concerning the variety of gendering and sexing practices which shape the terms on which innovation agents – as women and men with a variety of caring responsibilities and personal life commitments – are *differently* able to engage in the relational networks of learning and communities of practice long theorised as underpinning regional advantage. And all this in ways that necessarily make those regional learning dynamics much more complex than we have previously recognised (and indeed been willing to recognise). In the spirit of further disrupting and 'unlearning' these and other 'economic universals' (Gibson-Graham 2006) 'through the eyes of labour' (Herod 1997), the hybrid analysis developed in this book might usefully be extended in a number of important directions.

First, it is vital that future work continues to challenge the narrow focus of earlier work-life debates through an expanded focus on men (see also Brandth and Kvande 2001; Doucet 2006; Gregory and Milner 2011). Despite a now massive back catalogue, the majority of work-life balance studies to date have been concerned with how to flexibilise women's paid employment around an assumed female majority responsibility for childcare (Emslie and Hunt 2009; Burnett et al. 2010; Perrons et al. 2010). Yet as documented in Chapter 5, increasing numbers of Silicon Cowboys are hanging up their spurs and undertaking significant caring responsibilities outside of paid work, which precludes their ability (and willingness) to undertake sustained long hours working on behalf of their employers, and also generating significant work-life conflicts in the process. Not simply restricted to male technology workers in Dublin and Cambridge, the work-lives of these men need to be situated more broadly against (slowly!) shifting societal expectations around sharing unwaged carework at home, increased social acceptance of female-breadwinner households, rising childcare costs, the increasing dominance of feminised service sector employment in which masculinity is often seen as a labour market disadvantage, and a greater range of national policy entitlements promoting increased male uptake of care alongside paid employment.[1] This enlarged research focus is crucial because 'gender equity is only likely to be achieved if ... men become more "like women", combining the work of both employment and caregiving in their day to day lives' (Crompton 2006: 17). However, as was my experience in this study, it remains very challenging to encourage men to take part in work-life research in the first place, given powerful public perceptions of WLB as a 'female-only problem' (and within that, as the assumed narrow preserve of working mothers). A core component of this agenda will necessarily involve the development of innovative new methodological strategies to better engage men in WLB business case research. Future work should also include a targeted focus on employer-provided work-life arrangements that do more than simply *accept* the accommodation of working time to existing unequal gendered divisions of labour versus more radical provisions that *challenge* the models of male full-time employment/female primary carer by supporting the reallocation of childcare and reproductive tasks.

Second, in extending the analysis developed in this book, it is important that future research examines the positive impacts of similar bundles of WLB arrangements on learning and innovation outcomes in other knowledge-intensive sectors beyond IT. Other sectors in which workers have also been identified as regularly working long hours and struggling with a lack of work-life balance include law, banking and management consultancy.[2] Indeed, the potential for extending this multi-sectoral analysis even further is underlined by the dominant patterns of WLB provision evident amongst the 150 IT employers featured in this study, most commonly flextime, homeworking and part-time working, whose learning effects form the core focus of Chapter 6. Significantly, these are the same three most commonly provided WLB arrangements also evident in the UK's Fourth National Work-Life Balance Survey (2011), based on a much larger sample of 2,767 workers across multiple sectors. Future comparative work might also explore alternative strategies for work-life reconciliation amongst other digital workers in the so-called 'sharing economy' (or 'peer-to-peer economy'), defined in terms of the use of the internet as a means for unbundling production from employment, to enable digital labour to make money from unused human and physical assets – everything from cars and bicycles to housing, workplaces, food, household items, time or expertise (Rifkin 2014). Key questions emerge around the potential of the 'sharing economy' for having an empowering and/or liberating effect at a time when more and more people find it increasingly difficult to meet the demands of more formal, traditional work environments. Likewise the extent to which 'sharing work' disrupts stubborn gender inequalities in the household distribution of paid work and caring versus reinforcing those gender inequalities amongst so-called 'prosumers' (for whom traditional distinctions between producers and consumers are somewhat blurred). And while previous research has explored the work-lives of independent contractors (e.g. Sayah 2013), future analyses of work-life and learning amongst prosumers are faced with new conceptual difficulties, given the way in which some forms of online activity (e.g. uploading images, data and/or reviews to social media which are then used by companies for profit) do not neatly fit either established concepts of paid work or leisure, prompting the development of the hybrid concept of 'play-bour'. Likewise the additional analytical difficulties of dealing with employer-provided work-life arrangements, given that 'sharing work' blurs the core conceptual distinction between employ*er* and employ*ee*.

And third, it will be important for geographers to extend this analysis to other regional economies to gain a deeper understanding of the role of different local, regional and national legislative frameworks in shaping the impacts of work-life provision on everyday practices of learning and innovation in different place contexts. This work should consider how gendered patterns of cross-firm embodied knowledge (im)mobility prompted by work-life considerations vary between regions and cities with different average costs of living, transport costs, congestion, urban sprawl, average house prices, reliability of public transport, indices of liveability and rates of owner occupation. Likewise, what is the role of different national gender welfare regimes and policy frameworks in fostering more systemic and progressive possibilities for socially inclusive regional learning (especially in Scandinavian high-tech regions located in social democratic welfare regimes) by means of a reduced onus on employers to fill the WLB provision gap? As part of this geographical extended case comparison, it is also imperative that future research expands the focus of 'mainstream' analysis beyond advanced capitalist core economies to those in the Global South. This is important in order that we begin to provincialise 'universal' notions of the work-life balance metaphor which originated in the Western context (Lewis et al. 2007), and give voice to a wider variety of situated WLB experiences and knowledges. Examples of work-life balance research done outside the empirical heartlands of the UK, USA, Western Europe and Australia include work-life studies carried out in India, Pakistan, Hong Kong and Japan (e.g. Noor and Maad 2009; Wong and Ko 2009; Chandra 2012). But this work is too often characterised by tendencies to apply 'mainstream' WLB theory (developed in the Western empirical heartlands) to new cases, and to test other countries against an *a priori* WLB benchmark rooted in the West; rather than an opportunity to build genuinely new WLB theory from Southern empirical start-points. Key research questions remain, therefore, about what work-life balance looks like for different groups of workers in different countries in the Global South – and this with an analytical gaze that extends far beyond those workers employed at 'the other ends' of global service chains, servicing Western clients in real time through international chains of 'sub-contracted capitalism' (Wills 2009) emanating from the West. Similarly, to what extent are the peculiarly Western forms of work-life balance offshored by Western multinationals, subsequently disrupted and contested by offshore workers? (In other words, how are they 'geographically reconfigured' through the process of offshoring itself?) And what are the geographical possibilities for WLB/learning synergies in the informal sector, for workers in the Global South engaged in work that is not formally contractualised or recognised by the state? Much remains to be done.

Spreading the Word (in a language that can be heard)

Scholars have articulated repeated calls for enhanced employer assistance to enable workers to reconcile work, home and family, framed in terms of moral obligation, social justice and equality of opportunity (e.g. Connell 2005; Lewis and Haas 2005). However, such claims often have limited currency within the HR practitioner community, amongst the very managers tasked with convincing their senior colleagues to meet the costs of implementing more comprehensive suites of work-life provision (and who reminded me that they 'are not here to provide WLB because it is a nice thing to do, but because it affects the bottom line'). Likewise, the obvious amusement of the government minister and other members of the UK Department for Business Innovation and Skills who brushed away my apparently 'naïve' suggestion at a WLB roundtable lunch that government simply legislate to legally require that employers offer more comprehensive suites of WLB provision as standard. (A lunch that was also conveniently sponsored by a large well-known UK telecoms provider.) And also the members of the Girl Geek Dinners IT networking forum, who at their organisation's third anniversary dinner at Google HQ in London Victoria articulated their multiple experiences of either having their requests for 'alternative working' turned down on the basis of its perceived costs to their employers, or who else were too worried to make use of such arrangements in practice for fear of negative repercussions for their career development. Here, then, lies the wider relevance of the WLB dual agenda which has framed the analysis developed in this book, concerned to identify mutual gains for workers and firms in tandem, in which the gendered needs, feelings and preferences of workers as citizens and parents are analysed alongside the economic requirements of the firm. The aim is to construct a more socially progressive set of business case arguments, whilst also recognising the pragmatic need to frame calls for socially progressive corporate change in a language that managers (and shareholders) can hear (Froud et al. 2000).

That said, it is also important to recognise the pragmatic approach taken in *Work-Life Advantage* as one part of a broader and multi-scalar set of strategies for effecting positive change for workers and their families. This includes feminist calls for more progressive and radical work-life provision by employers framed around an ethical responsibility of government and employers to care; and the need for a wider socio-cultural shift that challenges masculinist career models rooted in presenteeism and long hours working by encouraging more men to assume an equal share of everyday childcare and social reproductive tasks. The aim, then, is to challenge a stubborn cultural asymmetry: while it has become more socially acceptable for women to engage in paid work, it has not become equally socially acceptable for men to reduce their commitment to paid work. Thus for Sirianni and Negrey (2000), the challenge is such that:

> The full equity potential of flexible work options requires generalising them to the point where alternative time cycles for achievement and recognition over the lifecourse, and differential time commitments to work in the market at any particular time, are viewed as genuinely legitimate. (p. 72)

Indeed, as one reviewer of this manuscript argued, the politics of welfare states are *the* most decisive factor in explaining the effects of flexible work and new work practices on female versus male workers, and on mothers versus fathers; and hence that 'looking for employer-initiated WLB arrangements is likely to be cosmetic compared to the transformative potential of welfare policies relating to gender and work'.

The dual agenda developed in this book seeks, therefore, to demonstrate empirically the *genuine legitimacy* of employer-provided WLB arrangements for regional learning and innovation. It is also consistent with the core terms of national WLB policy debate in the UK and Ireland over the last two decades, which remain stubbornly centred on the WLB business case. In the context of an ongoing neoliberal government refusal to intervene in firms' right to manage, there remains the need to frame calls for progressive corporate change in a language that managers (and shareholders) can hear. At the time of writing, these findings have been presented to academic audiences in the UK, Ireland, Sweden, Norway, Finland, Belgium and the USA; to industry audiences through Girl Geek Dinners, Women in Science and Engineering (WISE) and Women In Technology networking events and the Digital Shoreditch Festival; to representatives of the Trades Union Congress, Irish Congress of Trade Unions, Amicus, UNISON and UNITE; to members of Ireland's Equality Authority and the UK's Equality and Human Rights Commission; and used as part of a UK government's joint Equalities Office and Business, Innovation and Skills initiative to identify the evidence base for demonstrating the business benefits of equality and diversity. The results have also been disseminated internationally through the policy, practitioner and consultant networks developed through the Rosabeth Moss Kanter Work-Family Research Network. But much more remains to be done, and this book forms a key part of that dissemination process.

Postscript: Bringing It Home? From Work-Life Geographies to Geographers' Work-Lives

On my way out the door, I would like to bring the book full circle, back onto economic geographers themselves. My analysis shows that work-life conflict reduces the effectiveness of knowledge workers to do their jobs efficiently. It also shows that providing (and encouraging workers' use of) alternative working arrangements designed to enable workers to achieve a better balance between the demands of paid work, home and family also offers multiple learning advantages to employers. In portraying the kinds of work-life demands on IT workers as documented in Chapter 5 – including responsiveness to the demands of fee-paying customers, pervasive evening and weekend working, demands for overseas travel, pressures to network with peers after hours, increasing workloads – the similarities with academic work-lives are striking. So, too, the kinds of everyday coping mechanisms used by academic colleagues to muddle through (but which invariably short-change family), and the negative effects of this on patterns of career progression amongst female colleagues with children (the academic Mommy Track), and indeed among some male colleagues (the academic Daddy Track?). Likewise, the gendered constraints on cross-firm job-to-job mobility as documented in Chapter 7, as a function of workers' concerns to avoid disruptions (to delicately balanced commuting patterns combining nursery, school and a partner's commute; redistribution of domestic duties; possible relocation of home and school; and disruptions to established networks of family and community support) are also very apparent within academia, and extend well beyond dual-career academic couples. In short, I could have been writing about so many of my academic colleagues.

These patterns of work-life conflict amongst academics emerge from the demands of juggling multiple core activities of research, teaching, administration and student mentoring with extra-curricular responsibilities and commitments around personal life and family. These work-life conflicts form the focus of a recent book by US-based academics Maggie Berg and Barbara Seeber (2016), *The Slow Professor: Challenging the Culture of Speed in the Academy*, which calls for a slow-down in the pace of academic life through collective reform (rather than piecemeal individual fixes), and to 'take back the university'. Here in the UK, the combined pressures of the Research Excellence Framework, National Student Survey, Research Impact Agenda, Teaching Excellence Framework and Vice Chancellors' concerns to rise up through multiple university ranking tables are manifest through increased instances of work-related stress amongst academics, and motivate a major campaign by the UK's University and College Union (UCU) to improve work-life balance in UK academia. Likewise the growth of the UK's Athena SWAN Charter initiative, concerned to promote the career advancement of women in STEM subjects (and recently expanded to include humanities and social science), is also revealing in relation to the lived realities of work-life conflict amongst female academics. Through my involvement in four separate Athena SWAN bids to date, our cross-faculty consultations have documented multiple instances of female colleagues on reduced hours contracts but who maintain full hours workloads; a widespread inability and/or unwillingness to use their full holiday entitlement because the pressure to publish precludes this; and recent returners from maternity leave who, despite clear provision in the UK's Research Excellence Framework (REF) assessment exercise for a pro rata reduction in the required portfolio of four papers, are still asked to submit four so as not to lessen the size and quality of departmental submissions 'unnecessarily'. Likewise the absurdity of academic staff time use surveys which continue to cap the working week at 35 hours (less than half of the real figure for many colleagues). Also revealing are the kinds of responses by university employers. In Chapter 5, I identified IT workers' concerns around 'fruit-bowl-type' employer responses that target the *symptoms* of work-life conflict rather than deal with their fundamental *causes*. Sadly, such responses are not limited to technology employers. In one London-based Russell Group university, the managerial response to a 2015 staff survey that identified significant problems of work-life conflict amongst academic staff was to provide opportunities for colleagues to purchase body massages. Similarly, colleagues at another London-based Russell Group university were encouraged to destress by taking up the opportunity to stroke a guide dog puppy. Seriously.

My suggestion, then, is that employer scepticism around the so-called business case for work-life balance is far from restricted to technology employers. Rather, we too have a way to go in academia to recognise and endorse the multiple learning advantages that accrue to workers and employers alike through the provision and take-up of 'alternative' and 'non-traditional' working arrangements that better enable workers to reconcile competing demands and responsibilities of paid work, home and family. The key findings of this book around a more progressive WLB dual agenda suggest that we may well benefit from a rethink.

Notes

1 In the UK, those entitlements include shared parental leave and the recent extension of UK statutory paternity leave entitlement towards a Scandinavian model of shared childcare, and the greater acknowledgement given to fathers' caring roles in the UK Government Consultation on Modern Workplaces (BIS 2010)).

2 It is also important that future work broadens its focus beyond white collar professional work-lives to explore their (dis)connections to the work-lives of low-end service workers. As documented by colleagues at QMUL through the *Global Cities at Work* project, workers' everyday experiences of juggling three part-time, low-paid jobs (particularly in contract cleaning, home care, hospitality and food processing) give rise to rather different lived realities of work-life 'balance' than those documented here amongst white collar professional workers in predominantly full-time permanent positions (see Wills et al. 2010).

References

Acas 2009. *Flexible Working and Work-Life Balance*. London: Acas.

Acker, J. 1998. The future of 'gender and organizations': connections and boundaries. *Gender, Work and Organization* 5(45): 195–206.

Ackers, L. 2004. Managing relationships in peripatetic careers: scientific mobility in the European Union. *Women's Studies International Forum* 27(3): 189–201.

Ackroyd, S., Glover, I. and Currie, W. 2000. The triumph of hierarchies over markets: information system specialists in the current context. In Glover, I. and Hughes, M. (eds.) *Professions at Bay*. Aldershot: Avebury.

Agrawal, A., Cockburn, I. and McHale, J. 2006. Gone but not forgotten: knowledge flows, labor mobility, and enduring social relationships. *Journal of Economic Geography* 6: 571–591.

Ahuja, M.K., Chudoba, K.M. and Kacmar, C.J. 2007. IT road warriors: balancing work-family conflict, job autonomy and work overload to mitigate turnover intentions. *MIS Quarterly* 31(1): 1–17.

Allen, T.D. and Armstrong, J. 2006. Further examination of the link between work-family conflict and physical health. *American Behavioural Scientist* 49(9): 1204–1221.

Allen, T.D., Herst, D.E.L., Bruck, C.S. and Sutton, M. 2000. Consequences associated with work-family conflict: a review and agenda for future research. *Journal of Occupational Health Psychology* 5: 278–308.

Almeida, P. and Kogut, B. 1999. Localization of knowledge and the mobility of engineers in regional networks. *Management Science* 45(7): 905–917.

Amabile, T.M. 1996. *Creativity in Context*. Oxford: Westview Press.

Amabile, T.M. and Conti, R. 1997. Environmental determinants of work motivation, creativity, and innovation. In Garud, R., Nayyar, P.R. and Shapira, Z.B. (eds.) *Technological Innovation*. New York: Cambridge University Press, 111–125.

Work-Life Advantage: Sustaining Regional Learning and Innovation, First Edition. Al James.
© 2018 John Wiley & Sons Ltd. Published 2018 by John Wiley & Sons Ltd.

Amabile, T.M., Hadley, C.N. and Kramer, S.J. 2002. Creativity under the gun. *Harvard Business Review* 80(8): 52–61.

Amin, A. and Cohendet, P. 2004. *Architectures of Knowledge*. Oxford: Oxford University Press.

Anderson, S.E., Coffey, B.S. and Byerly, R.T. 2002. Formal organizational initiatives and informal workplace practices: links to work-life conflict and job-related outcomes. *Journal of Management* 28(6): 787–810.

Andres, H.P. 2002. A comparison of face-to-face and virtual software development teams. *Team Performance Management* 8(1/2): 39–48.

Arora, A., Gambardella, A. and Torrisi, S. 2001. *In the Footsteps of Silicon Valley? Indian and Irish Software in the International Division of Labour*. Stanford University, CA: Stanford Institute of Economic Policy Research Discussion Paper 00-41.

Arthur, M.B. and Rousseau, D.M. 1996. *The Boundaryless Career*. Oxford: Oxford University Press.

Asheim, B.T., Boschma, R. and Cooke, P. 2011. Constructing regional advantage: platform policies based on related variety and differentiated knowledge bases. *Regional Studies* 45(7): 893–904.

Asheim, B.T., Coenen, L. and Vang, J. 2007. Face-to-face, buzz, and knowledge bases: sociospatial implications for learning, innovation, and innovation policy. *Environment and Planning C* 25(5): 655–670.

Audretsch, D.B. and Feldman, M.P. 2004. Knowledge spillovers and the geography of innovation. In Henderson, V. and Thisse, J. (eds.) *Handbook of Urban and Regional Economics*. Amsterdam: Elsevier, 2713–2739.

Auerbach, J. 1990. Employer-supported child care as a women-responsive policy. *Journal of Family Issues* 11(4): 384–400.

Bacik, I. and Drew, E. 2006. Struggling with juggling. *Women's Studies International Forum* 29: 136–146.

Bailey, A.J., Blake, M.K. and Cooke, T.J. 2004. Migration, care, and the linked lives of dual-earner households. *Environment and Planning A* 36(9): 1617–1632.

Bailyn, L. 1985. Autonomy in the industrial R&D laboratory. *Human Resource Management* 24: 129–146.

Bailyn, L. 1993. *Breaking the Mold*. New York: The Free Press.

Bailyn, L. 1997. The impact of corporate culture on work-family integration. In Parasuraman, S. and Greenhaus, J.H. (eds.) *Integrating Work and Family: Challenges and Choices for a Changing World*. Westport, CT: Quorum Books, 209–219.

Bailyn, L., Fletcher, J.K. and Kolb, D. 1997. Unexpected connections: considering employees' personal lives can revitalise your business. *Sloan Management Review* 38: 11–19.

Baker, M. and Milligan, K. 2008. How does job-protected maternity leave affect mothers' employment? *Journal of Labor Economics* 26(4): 655–691.

Bakker, I. and Gill, S. (eds.) 2003. *Power, Production and Social Reproduction*. Basingstoke: Palgrave Macmillan.

Balbo, L. 1987. Crazy quilts: rethinking the welfare state debate from a woman's point of view. In Sassoon, A.S. (ed.) *Women and the State*. London: Hutchinson, 45–71.

Baltes, B.B., Briggs, T.E., Huff, J.W., Wright, J.A. and Neuman, G.A. 1999. Flexible and compressed workweek schedules: a meta-analysis of their effects on work-related criteria. *Journal of Applied Psychology* 84(4): 496–513.

Barnes, T.J. 1987. Homo economicus, physical metaphors, and universal models in economic geography. *The Canadian Geographer* 31(4): 299–308.

Barnes, T.J. and Sheppard, E. 2010. 'Nothing includes everything': towards engaged pluralism in Anglophone economic geography. *Progress in Human Geography* 34(2): 193–214.

Bathelt, H. 2008. Knowledge-based clusters: regional multiplier models and the role of 'buzz' and 'pipelines'. In Karlsson, C. (ed.) *Handbook of Research on Cluster Theory*. Cheltenham: Edward Elgar, 78–92.

Bathelt, H. and Cohendet, P. 2014. The creation of knowledge: local building, global accessing and economic development – toward an agenda. *Journal of Economic Geography* 14: 869–882.

Bathelt, H., Malmberg, A. and Maskell, P. 2004. Clusters and knowledge: local buzz, global pipelines and knowledge creation. *Progress in Human Geography* 28(1): 31–56.

Batt, R. and Valcour, P.M. 2003. Human resources practices as predictors of work-family outcomes and employee turnover. *Industrial Relations* 42(2): 189–220.

Beauregard, T.A. and Henry, L.C. 2009. Making the link between work-life balance practices and organizational performance. *Human Resource Management Review* 19: 9–22.

Beechey, V. and Perkins, T. 1987. *A Matter of Hours*. Minneapolis: University of Minnesota Press.

Benería, L. 2012. Gender and the social construction of markets. In Van Staveren, I., Elson, D., Grown, C. and Cagatay, N. (eds.) *Feminist Economics of Trade*. Abingdon: Routledge, 13–32.

Benn, M. 2002. Time is money, and we want both. *The Guardian*, 26 September 2002, 18.

Benner, C. 2002. *Work in the New Economy*. Oxford: Blackwell.

Benner, C. 2003. Learning communities in a learning region: the soft infrastructure of cross-firm learning communities in Silicon Valley. *Environment and Planning A* 35(10): 1809–1830.

Benner, C. and Pastor, M. 2012. *Just Growth*. London: Routledge.

Benner, C. and Pastor, M. 2015. *Equity, Growth, and Community*. Oakland, CA: University of California Press.

Benner, C., Leete, L. and Pastor, M. 2007. *Staircases or Treadmills?* New York: Russell Sage Foundation.

Berg, M. and Seeber, B. 2016. *The Slow Professor: Challenging the Culture of Speed in the Academy*. Toronto: University of Toronto Press.

Berik, G., Rodgers, Y.V.D.M. and Seguino, S. 2009. Feminist economics of inequality, development, and growth. *Feminist Economics* 15(3): 1–33.

Bevan, S., Dench, S., Tamkin, P. and Cummings, J. 1999. *Family-Friendly Employment: The Business Case*. DfEE Research Report.

Bevan, S., Kettley, P. and Patch, A. 1997. *Who Cares? The Business Benefits of Carer-Friendly Employment Policies*. Institute for Employment Studies, Report 330.

Bianchi, S.M. 2000. Maternal employment and time with children: dramatic change or surprising continuity? *Demography* 37(4): 401–414.

Bienkowska, D., Lundmark, M. and Malmberg, A. 2011. Brain circulation and flexible adjustment: labour mobility as a cluster advantage. *Geografiska Annaler B* 93(1): 21–39.

BIS 2010. *Consultation on Modern Workplaces*. London: Department for Business, Innovation and Skills.

Blake, M. and Hanson, S. 2005. Rethinking innovation: context and gender. *Environment and Planning A* 37(4): 681–701.

Blau, F.D., Ferber, M.A. and Winkler, A.E. 2002. *The Economics of Women, Men and Work.* Upper Saddle River, NJ: Prentice Hall.

Block, F. 1990. *Postindustrial Possibilities.* Berkeley, CA: University of California Press.

Bloom, N., Kretschmer, T. and Van Reenen, J. 2011. Are family-friendly workplace practices a valuable firm resource? *Strategic Management Journal* 32(4): 343–367.

Bloomberg 2013. Silicon Valley, where mass transit goes to die. *Bloomberg Business*, 12 February.

Bolger, N., DeLongis, A., Kessler, R. and Wethington, E. 1989. The contagion of stress across multiple roles. *Journal of Marriage and the Family* 51: 175–183.

Bond, J.T. and Galinsky, E. 2006. *How Can Employers Increase the Productivity and Retention of Entry-Level, Hourly Employees?* New York: Families and Work Institute.

Bonney, N. 2005. Overworked Britons? Part-time work and work-life balance. *Work, Employment and Society* 19: 391–401.

Bordo, S. 1990. Feminism, postmodernism, and gender scepticism. In Nicholson, L. (ed.) *Feminism/Postmodernism.* London: Routledge, 133–156.

Boschma, R. and Kloosterman, R.C. (eds.) 2005. *Learning from Clusters.* Dordrecht: Springer.

Boschma, R., Eriksson, R. and Lindgren, U. 2009. How does labour mobility affect the performance of plants? *Journal of Economic Geography* 9: 169–190.

Boschma, R., Eriksson, R.H. and Lindgren, U. 2014. Labour market externalities and regional growth in Sweden. *Regional Studies* 48(10): 1669–1690.

Boschma, R., Iammarino, S. and Steinmueller, E. 2013. Editorial: Geography, skills and technological change. *Regional Studies* 47(10): 1615–1617.

Boucher, G. and Collins, G. 2005. *The New World of Work.* Dublin: The Liffey Press.

Bowlby, S. 2012. Recognising the time-space dimensions of care: caringscapes and carescapes. *Environment and Planning A* 44: 2101–2118.

Bowlby, S., McKie, L., Gregory, S. and Macpherson, I. 2010. *Interdependency and Care Over the Lifecourse.* Abingdon: Routledge.

Boyer, I. 1993. *Flexible Working for Managers.* London: Chartered Institute of Management Accountants.

Boyer, S.L., Maertz, C.P. Jr, Pearso, A.W. and Keough, S. 2003. Work-family conflict: a model of linkages between work and family domain variables and turnover intentions. *Journal of Managerial Issues* 15: 175–190.

Brandon, P.D. and Temple, J.B. 2007. Family provisions at the workplace and their relationship to absenteeism, retention, and productivity of workers. *Australian Journal of Social Issues* 42(4): 447–460.

Brandth, B. and Kvande, E. 2001. Flexible work and flexible fathers. *Work, Employment and Society* 15(2): 251–267.

Breathnach, P. 1998. Exploring the 'Celtic Tiger' phenomenon: causes and consequences of Ireland's economic miracle. *European Urban and Regional Studies* 5(4): 305–316.

Breschi, S. and Lissoni, F., 2001. Knowledge spillovers and local innovation systems: a critical survey. *Industrial and Corporate Change* 10(4): 975–1005.

Breschi, S. and Lissoni, F. 2009. Mobility of skilled workers and co-invention networks: an anatomy of localized knowledge flows. *Journal of Economic Geography* 9: 439–468.

Breznitz, D. 2012. Ideas, structure, state action and economic growth: rethinking the Irish miracle. *Review of International Political Economy* 19(1): 87–113.

Bristow, G. 2005. Everyone's a winner: problematising the discourse of regional competitiveness. *Journal of Economic Geography* 5(3): 285–304.

Brodie, J. 1994. Shifting the boundaries: gender and the politics of restructuring. In Bakker, I. (ed.) *The Strategic Silence: Gender and Economic Policy*. London: Zed Press, 46–60.

Bronson, P. 1999. *The Nudist on the Late Shift*. New York: Random House.

Brown, J.S. and Duguid, P. 2000. Organizational learning and communities of practice: toward a unified view of working, learning and innovation. In Lesser, E.L., Fontaine, M.A. and Slusher, J.A. (eds.) *Knowledge and Communities*. Boston: Butterworth Heinemann, 99–121.

Bunnell, T.G. and Coe, N.M. 2001. Spaces and scales of innovation. *Progress in Human Geography* 25(4): 569–589.

Bunting, M. 2005. *Willing Slaves*. London: HarperCollins.

Burchell, B.J., Day, D., Hudson, M., Ladipo, F., Mankelow, R., Nolan, J., Reed, H., Wichert, I. and Wilkinson, F. 1999. *Job Insecurity and Work Intensification*. York: Joseph Rowntree Foundation.

Burchell, B.J., Ladipo, D. and Wilkinson, F. (eds.) 2002. *Job Insecurity and Work Intensification*. London: Routledge.

Burke, R.J. 2005. Foreword. In Lewis, S. and Cooper, C. (eds.) *Work-Life Integration: Case Studies of Organisational Change*. Chichester: John Wiley & Sons, ix–xii.

Burnett, S.B., Coleman, L., Houlston, C. and Reynolds, J. 2012. *Happy Homes and Productive Workplaces*. London: Working Families.

Burnett, S.B., Gatrell, C., Cooper, C.L. and Sparrow, P. 2010. Well-balanced families? A gendered analysis of work-life balance policies and work family practices. *Gender in Management* 25(7): 534–549.

Burud, S.L., Aschbacher, P.R. and McCroskey, J. 1984. *Employer-Supported Child Care*. Boston: Auburn House.

Calisir, F., Gumussoy, C.A. and Iskin, I. 2011. Factors affecting intention to quit among IT professionals in Turkey. *Personnel Review* 40(4): 514–533.

Capello, R. 1999. Spatial transfers of knowledge in high technology milieux: learning versus collective learning processes. *Regional Studies* 33(4): 353–365.

Carnoy, M. 2002. *Sustaining the New Economy*. Cambridge, MA: Harvard University Press.

Casper, W.J. and Buffardi, L.C. 2004. Work-life benefits and job pursuit intentions: the role of anticipated organizational support. *Journal of Vocational Behavior* 65: 391–410.

Castree, N. (ed.) 2004. *Spaces of Work*. London: Sage.

Castree, N. 2007. Labour geography: a work in progress. *International Journal of Urban and Regional Research* 31(4): 853–862.

CBI 2009. *Employment Trends 2009: Work Patterns in the Recession*. London: Confederation of British Industry.

Chandra, V. 2012. Work–life balance: eastern and western perspectives. *International Journal of Human Resource Management* 23(5): 1040–1056.

Charles, N. and Harris, C. 2007. Continuity and change in work-life balance choices. *British Journal of Sociology* 58(2): 277–295.

Chee, M.W.L. 2015. Limitations on visual information processing in the sleep-deprived brain and their underlying mechanisms. *Current Opinion in Behavioral Sciences* 1: 56–63.

Christophers, B., Lave, R., Peck, J. and Werner, M. 2016. 'Economic Transformations', a new book series in critical economic geography. Posted to CritGeogForum, 29 April 2016.

Christopherson, S. and Clark, J. 2007. *Remaking Regional Economies*. Abingdon: Routledge.

CIPD 2000. *Getting the Right Work-Life Balance*. London: CIPD.

Clark, S. 2000. Work-family border theory: a new theory of work-life balance. *Human Relations* 53(6): 747–770.

Coe, N.M. and Lier, D.J. 2011. Constrained agency? Re-evaluating the geographies of labour. *Progress in Human Geography* 35(2): 211–233.

Coenen, M. and Kok, R.A. 2014. Workplace flexibility and new product development performance: the role of telework and flexible work schedules. *European Management Journal* 32(4): 564–576.

Cohen, W.M. and Levinthal, D.A. 1990. Absorptive capacity: a new perspective on learning and innovation. *Administrative Science Quarterly* 35: 128–152.

Cohendet, P., Grandadam, D., Simon, L. and Capdevila, I. 2014. Epistemic communities, localization and the dynamics of knowledge creation. *Journal of Economic Geography* 14: 929–954.

Collins, G. and Boucher, G. 2005. Irish neo-liberalism at work? In Boucher, G. and Collins, G. (eds.) *The New World of Work*. Dublin: Liffey Press, 1–15.

Collins, R. 1990. Changing conceptions in the sociology of the professions. In Burrage, M. and Torstendahl, R. (eds.) *The Formation of Professions*. London: Sage, 11–23.

Connell, R.W. 2005. A really good husband: work/life balance, gender equity and social change. *Australian Journal of Social Issues* 40(3): 369–383.

Connolly, S., Aldrich, M., O'Brien, M., Speight, S. and Poole, E. 2013. *Fathers and Work*. London: Modern Fatherhood Study.

Cooke, P., Asheim, B., Boschma, R., Martin, R., Schwartz, D. and Tödtling, F. (eds.) 2011. *Handbook of Regional Innovation and Growth*. Cheltenham: Edward Elgar.

Cooper, C., Dewe, P. and O'Driscoll, M. 2001. *Organizational Stress*. London: Sage.

Cooper, M. 2000. Being the 'go-to guy': fatherhood, masculinity and the organization of work in Silicon Valley. *Qualitative Sociology* 23(4): 379–405.

Cousins, C.R. and Tang, N. 2004. Working time and work and family conflict in the Netherlands, Sweden and the UK. *Work, Employment and Society* 18(3): 531–549.

Cowling/Work Foundation 2005. *Still At Work?* London: The Work Foundation.

Cringeley, R.X. 1996. *Accidental Empires*. London: Penguin Books.

Crompton, R. 2006. *Employment and the Family*. Cambridge: Cambridge University Press.

Crompton, R. and Brockmann, M. 2006. Class, gender and work-life articulation. In Perrons, D., Fagan, C., McDowell, L. and Ward, K. (eds.) *Gender Divisions and Working Time in the New Economy*. Cheltenham: Edward Elgar, 103–121.

Crompton, R. and Lyonette, C. 2006. Work-life 'balance' in Europe. *Acta Sociologica* 49(4): 379–393.

Crompton, R., Lewis, S. and Lyonette, C. (eds.) 2007. *Women, Men, Work and Family in Europe*. Basingstoke: Palgrave Macmillan.

Crone, M. 2002. *A Profile of the Irish Software Industry*. Belfast: Northern Ireland Economic Research Centre.

Cross, C. and Linehan, M. 2006. Barriers to advancing female careers in the high-tech sector: empirical evidence from Ireland. *Women in Management Review* 21(1): 28–39.

Crouter, A.C., Bumpus, M.F., Head, M.R. and McHale, S.M. 2001. Implications of over-work and overload for the quality of men's family relationships. *Journal of Marriage and Family* 63(2): 404–416.

Crump, B.J., Logan, K.A. and McIlroy, A. 2007. Does gender still matter? A study of the views of women in the ICT industry in New Zealand. *Gender, Work and Organization* 14(4): 349–370.

Cruz, S.C. and Teixeira, A.A. 2009. The evolution of the cluster literature: shedding light on the regional studies–regional science debate. *Regional Studies* 44(9): 1263–1288.

CSO 2003. *Quarterly National Household Survey – Childcare Fourth Quarter*. Dublin: Central Statistics Office.

Cullen, E. 2004. Unprecedented growth – but for whose benefit? *Feasta Review* 2: 9–43.

Dahl, M.S. 2002. Embedded knowledge flows through labor mobility in regional clusters in Denmark. Paper presented to the DRUID Summer Conference, Copenhagen, 6–8 June.

Dahl, M.S. 2004. Embodied knowledge diffusion, labor mobility and regional dynamics: do social factors limit the development potential of regions? Paper presented to the DRUID Summer Conference, Copenhagen, 14-16 June.

Dalton, D.R. and Mesch, D.J. 1990. The impact of flexible scheduling on employee attendance and turnover. *Administrative Science Quarterly* 35: 270–287.

Darshan, M.S., Raman, R., Rao, T.S., Ram, D. and Annigeri, B. 2013. A study on professional stress, depression and alcohol use among Indian IT professionals. *Indian Journal of Psychiatry* 55(1): 63–69.

Davies, K. and Mathieu, C. 2005. *Gender Inequality in the IT Sector in Sweden and Ireland*. Stockholm: National Institute for Working Life.

Deal, T.E. and Kennedy, A.A. 2000. *Corporate Cultures*. Cambridge, MA: Perseus.

DeBord, K., Canu, R.F and Kerpelman, J. 2000. Understanding a work–family fit for single parents moving from welfare to work. *Social Work* 45(4): 313–324.

De Bruin, A. and Dupuis, A. 2004. Work-life balance? Insights from non-standard work. *New Zealand Journal of Employment Relations* 29(1): 21–37.

Deery, M. 2008. Talent management, work-life balance and retention strategies. *International Journal of Contemporary Hospitality Management* 20(7): 792–806.

Dembe, A.E., Erickson, J.B., Delbos, R.G. and Banks, S.M. 2005. The impact of overtime and long work hours on occupational injuries and illnesses. *Occupational and Environmental Medicine* 62(9): 588–597.

Dembosky, A. 2013. Young men in Silicon Valley choosing kids over work. *Financial Times*, April 19.

Dermott, E. and Miller, T. 2015. More than the sum of its parts? Contemporary father-hood policy, practice and discourse. *Families, Relationships and Societies* 4(2): 183–195.

Dex, S. 2003. *Work and Family in the Twenty-First Century*. York, UK: Joseph Rowntree Foundation.

Dex, S. and Joshi, H.E. 1999. Careers and motherhood: policy for compatibility. *Cambridge Journal of Economics* 23(5): 641–659.

Dex, S. and Scheibl, F. 1999. Business performance and family-friendly policies. *Journal of General Management* 24(4): 22–37.

Dex, S. and Scheibl, F. 2001. Flexible and family-friendly working arrangements in SMEs: business cases. *British Journal of Industrial Relations* 39(3): 411–431.

Dex, S. and Smith, C. 2002. *The Nature and Pattern of Family-Friendly Employment Policies in Britain*. Bristol: Policy Press.

Dex, S., Smith, C. and Winter, S. 2001. *Effects of family-friendly policies on business performance.* Judge Institute of Management Studies, Research Papers in Management Studies, University of Cambridge.

Dex, S., Ward, K. and Joshi, H. 2008. Changes in women's occupations and occupational mobility over 25 years. In Scott, J., Dex, S. and Joshi, H. (eds.) *Women and Employment.* Cheltenham: Edward Elgar, 54–80.

DfEE 2000. *Creating a Work-Life Balance: A Good Practice Guide for Employers.* London: Department for Education and Employment.

Díaz-García, C., González-Moreno, A. and Sáez-Martínez, F.J. 2013. Gender diversity within R&D teams: its impact on radicalness of innovation. *Innovation* 15(2): 149–160.

DiBella, A.J., Nevis, E.C. and Gould, J.M. 1996. Organizational learning style as a core capability. In Moingeon, B. and Edmondson, A. (eds.) *Organizational Learning and Competitive Advantage.* London: Sage, 38–55.

Dixon, Z.P. 2009. *The Neoliberal Privatisation of Risk and Responsibility: The Case of Work-Life Balance and Flexible Working Practices.* Unpublished doctoral thesis, Department of Sociology, University of Birmingham.

Doherty, L. 2004. Work-life balance initiatives: implications for women. *Employee Relations* 26(4): 433–452.

Doogan, K. 2009. *New Capitalism? The Transformation of Work.* Cambridge: Polity Press.

Doucet, A. 2006. *Do Men Mother?* Toronto: University of Toronto Press.

Doussard, M. 2013. *Degraded Work.* Minneapolis: University of Minnesota Press.

Drago, R. and Hyatt, D. 2003. Symposium: the effect of work-family policies on employees and employers. *Industrial Relations* 42: 139–145.

Drew, E., Humphreys, P. and Murphy, C. 2002. *Off the Treadmill: Achieving Work/Life Balance.* Dublin: National Framework Committee for Family Friendly Policies.

DTI 2004. *Flexible Working in the IT Industry.* London: Department of Trade and Industry.

Dumelow, C., Littlejohns, P. and Griffiths, S. 2000. Relation between a career and family life for English hospital consultants. *British Medical Journal* 320(27): 1437–1440.

Durbin, S. 2011. Creating knowledge through networks: a gender perspective. *Gender, Work and Organization* 18(1): 90–112.

DWP 2010. *Flexible Working: Working for Families, Working for Business.* London: Department for Work and Pensions.

Eaton, S.C. 2003. If you can use them: flexibility policies, organizational commitment, and perceived performance. *Industrial Relations* 42(2): 145–167.

Eccleston, J. 2011. Recession causing some employees to shun flexible working. *Personnel Today*, 18 November.

EHRC 2009. *Working Better.* London: Equality and Human Rights Commission.

EHRC/BIS 2015. *Pregnancy and Maternity-Related Discrimination and Disadvantage.* London: Equality and Human Rights Commission with the Department for Business, Innovation and Skills.

Eikhof, D.R., Warhurst, C. and Haunschild, A. 2007. Introduction: What work? What life? What balance? Critical reflections on the work-life balance debate. *Employee Relations* 29(4): 325–333.

Eisenberger, R., Fasolo, P. and Davis-LaMastro, V. 1990. Perceived organizational support and employee diligence, commitment and innovation. *Journal of Applied Psychology* 75: 51–59.

Elder, G.H. 1994. Time, human agency, and social change: perspectives on the life course. *Social Psychology Quarterly* 57: 4–15.

Employers for Work-Life Balance 2005. *Business Case: The Business Benefits*. London: Employers for Work-Life Balance.

Emslie, C. and Hunt, K. 2009. 'Live to work' or 'work to live'? A qualitative study of gender and work-life balance among men and women in mid-life. *Gender, Work and Organization* 16(1): 151–172.

England, K. 1991. Gender relations and the spatial structure of the city. *Geoforum* 22(2): 135–147.

England, P. 1993. The separative self: androcentric bias in neoclassical assumptions. In Ferber, M.A. and Nelson, J.A. (eds.) *Beyond Economic Man*. London: University of Chicago Press, 37–53.

England, P. 2003. Separative and soluble selves: dichotomous thinking in economics. In Ferber, M.A. and Nelson, J.A (eds.) *Feminist Economics Today*. London: University of Chicago Press, 33–59.

England, K. and Lawson, V. 2005. Feminist analyses of work: rethinking the boundaries, gendering, and spatiality of work. In Nelson, L. and Seager, J. (eds.) *A Companion to Feminist Geography*. Oxford: Blackwell, 77–92.

English-Lueck, J.A. 2010. *Being and Well-being: Health and the Working Bodies of Silicon Valley*. Stanford, CA: Stanford University Press.

Eriksson, R. and Lindgren, U. 2009. Localized mobility clusters: impacts of labour market externalities on firm performance. *Journal of Economic Geography* 9: 33–53.

Eriksson, R. and Lindgren, U. 2011. What about people in evolutionary economic geography? Paper presented to the 3rd Global Conference on Economic Geography, Seoul, Korea, 28 June–2 July.

Esping-Andersen, G. 1990. *The Three Worlds of Welfare Capitalism*. Princeton: Princeton University Press.

Esping-Andersen, G. 1999. *Social Foundations of Postindustrial Economies*. Oxford: Oxford University Press.

Esping-Andersen, G. 2009. *The Incomplete Revolution*. Cambridge: Polity Press.

Ettlinger, N. 2003. Cultural economic geography and a relational and microspace approach to trusts, rationalities, networks, and change in collaborative workplaces. *Journal of Economic Geography* 3(2): 145–171.

EU-OSHA 2014. *Calculating the Cost of Work-Related Stress and Psychosocial Risks*. Luxembourg: European Agency for Safety and Health at Work.

Eurofound 2012. *European Quality of Life Survey 2012 Overview Report: Quality of Life in Europe, Impact of the Crisis*. Dublin: European Foundation for the Improvement of Living and Working Conditions.

European Commission 2006. *The Rationale of Motherhood Choices: Influence of Employment Conditions and of Public Policies*. Brussels: European Commission.

European Commission 2010a. *Europe 2020: A Strategy for Smart, Sustainable and Inclusive Growth*. Brussels: Publications Office of the European Union.

European Commission 2010b. *Women and ICT Status Report*. Brussels: Publications Office of the European Union.

Eurostat/European Union 2005. *Eurostat European Union Labour Force Survey 2004*. Luxembourg: Office for Official Publications of the European Communities.

Eurostat/European Union 2015. *Eurostat European Union Labour Force Survey 2014.* Luxembourg: Office for Official Publications of the European Communities.

Faggian, A. and McCann, P. 2006. Human capital flows and regional knowledge assets: a simultaneous equation approach. *Oxford Economic Papers* 58(3): 475–500.

Fahey, T. and Duffy, D. 2008. The housing boom. In Fahey, T., Russell, H. and Whelan, C.T. (eds.) *Quality of Life in Ireland.* Milton Keynes: Springer, 123–138.

Fahey, T., Russell, H. and Whelan, C.T. (eds.) 2008. *Quality of Life in Ireland.* Milton Keynes: Springer.

Fallick, B., Fleischman, C.A. and Rebitzer, J.B. 2004. Job-hopping in Silicon Valley: the micro-foundations of a high technology cluster. OECD working papers on labour markets, human capital and inequality.

Fallick, B., Fleischman, C.A. and Rebitzer, J.B. 2006. Job-hopping in Silicon Valley: the micro-foundations of a high tech cluster. *Review of Economics and Statistics* 88(3): 472–481.

Farrands, C. 1997. Interpretations of the diffusion and absorption of technology. In Talalay, M., Farrands, C. and Tooze, R. (eds.) *Technology, Culture and Competitiveness.* London: Routledge, 75–89.

Faulkner, W. 2000. Dualisms, hierarchies and gender in engineering. *Social Studies of Science* 30: 759–792.

Fawcett Society 2009. *Are Women Bearing the Burden of the Recession?* London: Fawcett Society.

Feldman, J.M. and Klofsten, M. 2000. Medium-sized firms and the limits to growth: a case study in the evolution of a spin-off firm. *European Planning Studies* 8(5): 631–650.

Felmlee, D.H. 1982. Women's job mobility processes within and between employers. *American Sociological Review* 47(1): 142–151.

Ferber, M.A. and Nelson, J.A. (eds.) 1993. *Beyond Economic Man.* London: University of Chicago Press.

Ferber, M.A. and Nelson, J.A. (eds.) 2003. *Feminist Economics Today.* London: University of Chicago Press.

Figart, D.M. and Mutari, E. 1998. Degendering work time in comparative perspective: alternative policy frameworks. *Review of Social Economy* 56(4): 460–480.

Firth, L., Mellor, D.J., Moore, K.A. and Loquet, C. 2004. How can managers reduce employee intention to quit? *Journal of Managerial Psychology* 19(2): 170–187.

Fischer, P.A. and Malmberg, G. 2001. Settled people don't move: on life course and (im-) mobility in Sweden. *International Journal of Population Geography* 7: 357–371.

Fleetwood, S. 2007. Why work-life balance now? *International Journal of Human Resource Management* 18(3): 387–400.

Florida, R. 2002. *The Rise of the Creative Class.* New York: Basic Books.

Folbre, N. 1994. *Who Pays for the Kids? Gender and the Structure of Constraint.* London: Routledge.

Folbre, N. and Nelson, J.A. 2000. For love or money – or both? *Journal of Economic Perspectives* 14(4): 123–140.

Forbes 2012. Silicon Valley: work-life balance is for losers, not closers. *Forbes Magazine*, 8 October.

Forbes 2013. What's the dark side of Silicon Valley? *Forbes Magazine*, 30 May.

Forret, M.L. and Dougherty, T.W. 2004. Networking behaviors and career outcomes: differences for men and women? *Journal of Organizational Behavior* 25: 419–437.

Forsyth, S. and Polzer-Debruyne, A. 2007. The organisational pay-offs for perceived work–life balance support. *Asia Pacific Journal of Human Resources* 45(1): 113–123.

Forum on the Workplace of the Future 2005. *Working to Our Advantage: A National Workplace Strategy*. Dublin: NCPP.

Frisén, A., Carlsso, J. and Wängqvist, M. 2014. 'Doesn't everyone want that? It's just a given': Swedish emerging adults' expectations on future parenthood and work/family priorities. *Journal of Adolescent Research* 29(1): 67–88.

Frolick, M.N., Wilkes, R.N. and Urwiler, R. 1993. Telecommuting as a workplace alternative. *Journal of Strategic Information Systems* 2: 206–222.

Frone, M.R., Barnes, G.M. and Farrell, M.P. 1994. Relationships of work-family conflict to substance abuse among employed mothers. *Journal of Marriage and Family* 56: 1019–1030.

Frone, M.R., Russell, M. and Cooper, M.L. 1997. Relation of work–family conflict to health outcomes: a four-year longitudinal study of employed parents. *Journal of Occupational and Organizational Psychology* 70(4): 325–335.

Froud, J., Haslam, C., Johal, S. and Williams, K. 2000. Restructuring for shareholder value and its implications for labour. *Cambridge Journal of Economics* 24(6): 771–797.

Fujigaki, Y. and Mori, K. 1997. Longitudinal study of work stress among information system professionals. *International Journal of Human-Computer Interaction* 9(4): 369–381.

Fujigaki, Y., Asakura, T. and Haratani, T. 1994. Work stress and depressive symptoms among Japanese information systems managers. *Industrial Health* 32(4): 231–238.

Galinsky, E. and Bond, J.T. 2009. *The Impact of the Recession on Employers*. New York: Families and Work Institute.

Galinsky, E. and Johnson, A. 1998. *Reframing the Business Case for Work-Life Initiatives*. New York: Families and Work Institute.

Galinsky, E., Bond, J.T. and Sakai, K. 2008. *2008 National Study of Employers*. New York: Family and Work Institute.

Gallagher, A. 2012. Neoliberal governmentality and the respatialisation of childcare in Ireland. *Geoforum* 43: 464–471.

Gambles, R., Lewis, S. and Rapoport, R. 2006. *The Myth of Work-Life Balance*. Chichester: John Wiley & Sons.

Gardiner, B., Martin, R. and Tyler, P. 2004. Competitiveness, productivity and economic growth across the European regions. *Regional Studies* 38(9): 1045–1067.

Gatrell, C.J. and Cooper, C.L. 2008. Work-life balance: working for whom? *European Journal of International Management* 2(1): 71–86.

Gershuny, J. 2003. *Changing Times*. Oxford: Oxford University Press.

Gershuny, J. 2005. Busyness as the badge of honour of the new superordinate working class. Institute for Social and Economic Research, Working Paper 2005-09. Colchester: University of Essex.

Gerson, K. 1993. *No Man's Land*. New York: Basic Books.

Gertler, M.S. 1997. The invention of regional culture. In Lee, R. and Wills, J. (eds.) *Geographies of Economies*. London: Arnold, 47–58.

Gertler, M.S. 2002. Technology, culture and social learning: regional and national institutions of governance. In Gertler, M.S. and Wolfe, D.A. (eds.) *Innovation and Social Learning*. Basingstoke: Palgrave MacMillan, 111–134.

Gertler, M. 2003. Tacit knowledge and the economic geography of context, or the indefinable tacitness of being (there). *Journal of Economic Geography* 3: 75–99.

Gertler, M.S. and Wolfe, D.A. (eds.) 2002. *Innovation and Social Learning*. Basingstoke: Palgrave Macmillan.

Ghoshal, L. and Passerini, K. 2006. Understanding gender discrimination in information systems: a review. In Trauth, E. (ed.) *Encyclopedia of Gender and IT*. Harrisburg, PA: Idea Group, 25–30.

Gibson-Graham, J.K. 1993. Waiting for the revolution: or how to smash capitalism while working at home in your spare time. *Rethinking Marxism* 6: 10–24.

Gibson-Graham, J.K. 1996. *The End of Capitalism (as we knew it)*. Oxford: Blackwell.

Gibson-Graham, J.K. 2006. *The End of Capitalism (as we knew it)*, 2nd edition. Minneapolis: University of Minnesota Press.

Gibson-Graham, J.K., Cameron, J. and Healy, S. 2013. *Take Back the Economy*. Minneapolis: University of Minnesota Press.

Glass, J.L. and Estes, S.B. 1997. The family responsive workplace. *Annual Review of Sociology* 23: 289–313.

Glass, J.L. and Finley, A. 2002. Coverage and effectiveness of family-responsive workplace policies. *Human Resource Management Review* 12: 313–337.

Glass, J.L. and Riley, L. 1998. Family responsive policies and employee retention following childbirth. *Social Forces* 76(4): 1401–1435.

Glynn, C., Steinberg, I. and McCartney, C. 2002. *Work-Life Balance: The Role of the Manager*. West Sussex: Roffey Park Institute.

Goff, S., Mount, M. and Jamison, R. 1990. Employer supported child care, work/family conflict and absenteeism: a field study. *Personnel Psychology* 43: 793–809.

Good, L.K., Page, T.J. and Young, C.E. 1996. Assessing hierarchical differences in job-related attitudes and turnover among retail managers. *Journal of the Academy of Marketing Science* 24: 148–156.

Gornick, J.C. and Meyers, M.K. 2003. *Families That Work*. New York: Russell Sage Foundation.

Grabher, G., Ibert, O. and Flohr, S. 2008. The neglected king: the customer in the new knowledge ecology of innovation. *Economic Geography* 84(3): 253–280.

Gray, M. and James, A. 2007. Connecting gender and economic competitiveness: lessons from Cambridge's high tech regional economy. *Environment and Planning A* 39(2): 417–436.

Gray, M. and Parker, E. 1998. Industrial change and regional development: the case of the US biotechnology and pharmaceutical industries. *Environment and Planning A* 30(10): 1757–1774.

Gray, M., Kurihara, T., Hommen, L. and Feldman, J. 2007. Networks of exclusion: job segmentation and gendered social networks in the knowledge economy. *Equality, Diversity and Inclusion* 26(2): 144–161.

Greco, L. 2005. Women and work in the Irish ICT sector. In Boucher, G. and Collins, G. (eds.) *The New World of Work*. Dublin: Liffey Press, 83–104.

Green, A.E. 1997. A question of compromise? Case study evidence on the location and mobility strategies of dual career households. *Regional Studies* 31(7): 641–657.

Green, F. and McIntosh, S. 2001. The intensification of work in Europe. *Labour Economics* 8: 291–308.

Greenblatt, E. 2002. Work-life balance: wisdom or whining. *Organisational Dynamics* 31(2): 177–193.

Greenhaus, J.H. and Beutell, N.J. 1985. Sources of conflict between work and family roles. *Academy of Management Review* 10(1): 76–88.

Greenhaus, J.H. and Singh, R. 2003. *Work–family linkages: a Sloan work and family encyclopedia entry*. Chestnut Hill, MA: Boston College.

Greenhaus, J.H., Collins, K.M. and Shaw, J.D. 2003. The relation between work–family balance and quality of life. *Journal of Vocational Behavior* 63(3): 510–531.

Greenhaus, J.H., Collins, K.M., Singh, R. and Parasuraman, S. 1997. Work and family influences on departure from public accounting. *Journal of Vocational Behavior* 50: 249–270.

Gregg, M. 2011. *Work's Intimacy*. Cambridge: Polity Press.

Gregory, A. and Milner, S. 2009. Work-life balance: a matter of choice? *Gender, Work and Organization* 16(1): 1–13.

Gregory, A. and Milner, S. 2011. Fathers and work-life balance in France and the UK: policy and practice. *International Journal of Sociology and Urban Policy* 31(1/2): 34–52.

Gregory, A., Milner, S. and Windebank, J. 2013. Work-life balance in times of economic crisis and austerity. *International Journal of Sociology and Social Policy* 33(9/10): 528–541.

Grimes, S. 2003. Ireland's emerging information economy: recent trends and future prospects. *Regional Studies* 37(1): 3–14.

Grover, S. and Crooker, K. 1995. Who appreciates family responsive human resource policies: the impact of family friendly policies on the organizational attachment of parents and non-parents. *Personnel Psychology* 48: 271–288.

Guardian, The 2013. Bank of America intern's death puts banks' working culture in spotlight. 21 August.

Guardian, The 2014. Let's do lunch. 23 June.

Guardian, The 2015a. Too many skipped lunches can be bad for your career. 26 March.

Guardian, The 2015b. You want a career and a baby? Who do you think you are? 24 February.

Guest, D.E. 2002. Perspectives on the study of work-life balance. *Social Science Information* 41(2): 255–279.

Haggett, P. 1995. *The Geographer's Art*. Oxford: Wiley Blackwell.

Halpern, D.F. 2005. How time-flexible work policies can reduce stress, improve health, and save money. *Stress and Health* 21(3): 157–168.

Han, S.K. and Moen, P. 1999. Work and family over time: a life course approach. *Annals of the American Academy of Political and Social Science* 562(1): 98–110.

Hannah, R. 1994. The trade-off between worker mobility and employer flexibility: recent evidence and implications. *Employee Benefits Journal* 19(2): 23–25.

Hanson, S. and Pratt, G. 1988. Reconceptualizing the links between home and work in urban geography. *Economic Geography* 64(4): 299–321.

Hanson, S. and Pratt, G. 1991. Job search and the occupational segregation of women. *Annals of the Association of American Geographers* 81(2): 229–253.

Hanson, S. and Pratt, G. 1995. *Gender, Work and Space*. London: Routledge.

Hardhill, I. 2002. *Gender, Migration, and the Dual Career Household*. London: Routledge.

Hardhill, I. and van Loon, J. 2006. Individualization and 'identity-risks' in dual-career households. In Perrons, D., Fagan, C., McDowell, L. and Ward, K. (eds.) *Gender Divisions and Working Time in the New Economy*. Cheltenham: Edward Elgar, 162–177.

Harding, S.G. 1986. *The Science Question in Feminism.* Ithaca, NY: Cornell University Press.

Harding, S.G. 1991. *Whose Science? Whose Knowledge? Thinking From Women's Lives.* Ithaca, NY: Cornell University Press.

Harding, S.G. (ed.) 2004. *The Feminist Standpoint Theory Reader.* New York: Routledge.

Hardy, K. 2010. (Sex) working class subjects: integrating sex workers in the Argentine labour movement. *International Labour and Working Class History* 77(1): 89–108.

Harvey, D. 1982. *The Limits to Capital.* Oxford: Blackwell.

Harvey, D. 2005. *A Brief History of Neoliberalism.* Oxford: Oxford University Press.

Hayward, B., Fong, B. and Thornton, A. 2007. *The Third Work-Life Balance Employer Survey: Main Findings.* London: Department for Business and Enterprise.

He, J., Butler, B.S. and King, W.R. 2007. Team cognition: development and evolution in software project teams. *Journal of Management Information Systems* 24(2): 261–292.

Healy, G. 2004. Work-life balance and family friendly policies – in whose interest? *Work, Employment and Society* 18(1): 219–223.

Hearst, M.O., Sevcik, S., Fulkerson, J.A., Pasch, K.E., Harnack, L.J. and Lytle, L.A. 2012. Stressed out and overcommitted! The relationships between time demands and family rules and parents' and their child's weight status. *Health Education and Behavior* 39(4): 446–454.

Heeks, R. and Nicholson, B. 2004. Software export success factors and strategies in 'follower' nations. *Competition and Change* 8(3): 267–303.

Heiland, F. and MacPherson, D.A. 2005. *Do Family-Friendly Firms Perform Better?* Working paper series, Florida State University Department of Economics.

Henry, N. and Pinch, S. 2000. Spatialising knowledge: placing the knowledge community of Motor Sport Valley. *Geoforum* 31: 191–208.

Henwood, F. 1993. Establishing gender perspectives on information technology. In Green, E., Owen, J. and Pain, D. (eds.) *Gendered by Design?* London: Taylor and Francis, 31–49.

Herod, A. 1997. From a geography of labor to a labor geography. *Antipode* 29(1): 1–31.

Herod, A. 2001. *Labor Geographies.* London: Guilford Press.

Herod, A., Rainnie, A. and McGrath-Champ, S. 2007. Working space: why incorporating the geographical is central to theorizing work and employment practices. *Work, Employment and Society* 21(2): 247–264.

Hewitt, P. 1993. *About Time.* London: IPPRI Rivers Orram Press.

Highsmith, J. 2002. *Agile Software Development Ecosystems.* Boston: Addison-Wesley Professional.

Hilbrecht, M., Shaw, S.M., Johnson, L.C. and Andrey, J. 2008. 'I'm home for the kids': contradictory implications for work-life balance of teleworking mothers. *Gender, Work and Organization* 15(5): 454–476.

Hobsbawm, J. 2009. *The See-Saw.* London: Atlantic Books.

Hochschild, A.R. 1983. *The Managed Heart.* Berkeley: University of California Press.

Hochschild, A.R. 1989. *The Second Shift.* New York: Viking.

Hochschild, A.R. 1997. *The Time Bind: When Work Becomes Home and Home Becomes Work.* New York: Metropolitan Books.

Hochschild, A.R. 2003. *The Commercialisation of Intimate Life.* Berkeley: University of California Press.

Hochschild, A.R. 2012. *The Outsourced Self.* New York: Metropolitan Books.

Hoegl, M. and Gemuenden, H.G. 2001. Teamwork quality and the success of innovative projects. *Organization Science* 12(4): 435–449.

Hogarth, T., Hasluck, C., Pierre, G., Winterbotham, M. and Vivian, D. 2001. *Work-Life Balance 2000.* Warwick: Institute for Employment Research. DfEE Research Report 249.

Hogarth, T., Owen, D., Gambin, L., Hasluck, C., Lyonette, C. and Casey, B. 2009. *The Equality Impacts of the Current Recession.* Equality and Human Rights Commission, Research Report 47.

Honeycutt, T.L. and Rosen, B. 1997. Family friendly human resource policies, salary levels, and salient identity as predictors of organizational attraction. *Journal of Vocational Behavior* 50: 271–290.

Hook, J. 2006. Care in context: men's unpaid work in 20 countries, 1965–2003. *American Sociological Review* 71: 639–660.

Hooker, L., Neathey, F., Casebourne, J. and Munro, M. 2011. *The Third Work-Life Balance Employee Survey.* London: Department for Business Innovation and Skills.

Horelli, L. 1997. Engendering evaluation of European regional development. *Evaluation* 3(4): 435–450.

Houston, D. (ed.) 2005. *Work-Life Balance in the Twenty-First Century.* New York: Palgrave.

Huber, F. 2012. On the role and interrelationship of spatial, social and cognitive proximity. *Regional Studies* 46(9): 1169–1182.

Hudson, M. 2002. Flexibility and the reorganisation of work. In Burchell, B.J., Ladipo, D. and Wilkinson, F. (eds.) *Job Insecurity and Work Intensification.* London: Routledge, 39–60.

Huhtala, H. and Parzefall, M.R. 2007. A review of employee well-being and innovativeness: an opportunity for a mutual benefit. *Creativity and Innovation Management* 16(3): 299–306.

Hui, K. 2014. The obstacles of female entrepreneurship in Silicon Valley. *Intersect: The Stanford Journal of Science, Technology and Society* 7(2): 1–12.

Human Resource Management 1996. The business argument for flexibility. *Human Resource Management,* May.

Hurst, J., Skinner, D. and Worrall, L. 2009. *The 24-7 Work-Life Balance Survey 2009.* Newcastle-under-Lyme, England: University of Keele Work/Life Balance Centre.

Huselid, M. 1995. The impact of human resource management practices on turnover, productivity, and corporate financial performance. *Academy of Management Journal* 38(3): 635–672.

Hyman, J. and Summers, J. 2004. Lacking balance? Work-life employment practices in the modern economy. *Personnel Review* 33(4): 418–429.

Hyman, J., Baldry, C., Scholarios, D. and Bunzel, D. 2003. Work–life imbalance in call centres and software development. *British Journal of Industrial Relations* 41(2): 215–239.

IBEC 2002. *Family-Friendly/Work-Life Balance.* Dublin: IBEC Research and Information Services.

ICTU 2005. *Family Friendly Working and Work-Life Balance Guidelines.* Dublin: Irish Congress of Trades Unions.

Independent, The 2015. Goldman Sachs analyst found dead hours after complaining to father of '100 hour weeks'. 3 June.

Jaffe, A.B., Trajtenberg, M. and Henderson, R. 1993. Geographic localization of knowledge spillovers as evidenced by patent citations. *Quarterly Journal of Economics* 108(3): 577–598.

James, A. 2005. Demystifying the role of culture in innovative regional economies. *Regional Studies* 39(9): 1197–1216.

James, A. 2008. Gendered geographies of high tech regional economies. *Geography Compass* 2 (1): 176–198.

James, A. 2011. Work-life (im)'balance' and its consequences for everyday learning and innovation in the New Economy: evidence from the Irish IT sector. *Gender, Place and Culture* 18(5): 655–684.

James, A. 2013. Work-life 'balance' and gendered (im)mobilities of knowledge and learning in regional economies. *Journal of Economic Geography* 14(3): 483–510.

James, A. 2014. Work-life 'balance', recession and the gendered limits to learning and innovation (or, why it pays employers to care). *Gender, Work and Organization* 21(3): 273–294.

Janssen, O., Van de Vliert, E. and West, M. 2004. The bright and dark sides of individual and group innovation. *Journal of Organizational Behavior* 25(2): 129–145.

Jarvis, H. 2005. *Work/Life City Limits*. Basingstoke: Palgrave Macmillan.

Jarvis, H. and Pratt, A. 2006. Bringing it all back home: the extensification and 'overflowing' of work. *Geoforum* 37: 331–339.

Johnson, J.H. 2004. Do long work hours contribute to divorce? *BE Journal of Economic Analysis and Policy* 4(1): 1–25.

Jones, A. 2003. *About Time for a Change*. London: The Work Foundation/Employers for Work-Life Balance.

Joo, B., McLean, G.N. and Yang, B. 2013. Creativity and human resource development: an integrative literature review and a conceptual framework for future research. *Human Resource Development Review* 12: 390.

Josten, E.J., Ng-A-Tham, J.E. and Thierry, H. 2003. The effects of extended workdays on fatigue, health, performance and satisfaction in nursing. *Journal of Advanced Nursing* 44(6): 643–652.

Kahn, W.A. 1990. Psychological conditions of personal engagement and disengagement at work. *Academy of Management Journal* 33: 692–724.

Kanter, R.M. 1977. *Work and Family in the United States*. New York: Russell Sage.

Kanter, R.M. 1983. *The Changemasters*. New York: Simon and Schuster.

Kanter, R.M. 1988. When a thousand flowers bloom: structural, collective, and social conditions for innovation in organizations. *Research in Organizational Behavior* 10: 169–211.

Karatepe, O.M. and Uludag, O. 2007. Conflict, exhaustion, and motivation: a study of frontline employees. *International Journal of Hospitality Management* 26(3): 645–665.

Katz, C. 2001. Vagabond capitalism and the necessity of social reproduction. *Antipode* 33(4): 709–728.

Keeble, D., Lawson, C., Moore, B. and Wilkinson, F. 1999. Collective learning processes, networking, and institutional thickness in the Cambridge region. *Regional Studies* 33(4): 319–332.

Kersley, B., Alpin, C., Forth, J., Bryson, A., Bewley, H., Dix, G. and Oxenbridge, S. 2006. *Inside the Workplace: Findings from the 2004 Workplace Employment Relations Survey*. London: Routledge.

Khallash, S. and Kruse, M. 2012. The future of work and work-life balance 2025. *Futures* 44(7): 678–686.

Kidder, T. 1981. *The Soul of a New Machine*. Boston: Little, Brown.

Kilkey, M., Perrons, D. and Plomien, A. 2012. *Gender, Migration and Domestic Work.* New York: Palgrave Macmillan.

Kingston, P. 1990. Illusions and ignorance about the family responsive workplace. *Journal of Family Planning Issues* 11: 438–454.

Kirby, P. 2002. *The Celtic Tiger in Distress.* Basingstoke: Palgrave.

Konrad, A.M. and Mangel, R. 2000. The impact of work-life programs on firm productivity. *Strategic Management Journal* 21: 1225–1237.

Kopelman, R.E., Greenhaus, J.H. and Connolly, T.F. 1983. A model of work, family, and interrole conflict: a construct validation study. *Organizational Behavior and Human Performance* 32(2): 198–215.

Kossek, E.E. and Nichol, V. 1992. Effects of on-site child care on employee attitudes and performance. *Personnel Psychology* 45: 485–509.

Kossek, E.E., Baltes, B.B. and Matthews, R.A. 2011. How work-family research can finally have an impact in organizations. *Industrial and Organizational Psychology* 4(3): 352–369.

Kossek, E.E., Lewis, S. and Hammer, L.B. 2010. Work-life initiatives and organizational change: overcoming mixed messages to move from the margin to the mainstream. *Human Relations* 63(1): 3–19.

Kratzer, J., Leenders, R.T.A.J. and van Engelen, J.M.L. 2004. Stimulating the potential: creative performance and communication in innovation teams. *Creativity and Innovation Management* 13(1): 63–71.

Krugman, P. 1991. Increasing returns and economic geography. *Journal of Political Economy* 99: 483–499.

Krugman, P. 1993. On the relationship between trade theory and location theory. *Review of International Economics* 1(2): 110–122.

Kunda, G. 2006. *Engineering Culture: Control and Commitment in a High-Tech Corporation.* Philadelphia: Temple University Press.

Kvande, E. 2009. Work-life balance for fathers in globalized knowledge work: some insights from the Norwegian context. *Gender, Work and Organization* 16(1): 58–62.

Kwan, M. 1999. Gender, the home-work link, and space-time patterns of nonemployment activities. *Economic Geography* 75(4): 370–394.

La Valle, I., Arthur, S., Millward, C., Scott, J. and Clayden, M. 2002. *Happy Families? Atypical Work and its Influence on Family Life.* Bristol: Policy Press.

Lave, J. and Wenger, E. 1991. *Situated Learning.* Cambridge: Cambridge University Press.

Lawson, C. and Lorenz, E. 1999. Collective learning, tacit knowledge and regional innovative capacity. *Regional Studies* 33: 305–317.

Lawson, V. 2007. Geographies of care and responsibility. *Annals of the Association of American Geographers* 97(1): 1–11.

Lawton-Smith, H. and Waters, R. 2005. Employment mobility in high-technology agglomerations: the cases of Oxfordshire and Cambridgeshire. *Area* 37(2): 189–198.

Lazzeretti, L., Sedita, S.R. and Caloffi, A. 2014. Founders and disseminators of cluster research. *Journal of Economic Geography* 14(1): 21–43.

Leighton, D. and Gregory, D. 2011. *Reinventing the Workplace.* London: DEMOS.

Lewis, J. 2009. *Work-Family Balance, Gender and Policy.* Cheltenham: Edward Elgar.

Lewis, J. 2010. Reflecting on impact, changes and continuities – Restructuring workplace cultures: the ultimate work-family challenge. *Gender in Management* 25(5): 348–354.

Lewis, J. and Campbell, M. 2007. What's in a name? 'Work and family' or 'work and life' balance policies in the UK since 1997 and the implications for the pursuit of gender equality. *Social Policy and Administration* 42(5): 524–541.

Lewis, S. and Cooper, C. 1999. The work-family research agenda in changing contexts. *Journal of Occupational Health Psychology* 4(4): 382–393.

Lewis, S. and Cooper, C. 2005. *Work-Life Integration: Case Studies of Organisational Change.* Chichester: John Wiley & Sons.

Lewis, S. and Haas, L. 2005. Work-life integration and social policy: a social justice theory and gender equity approach to work and family. In Kossek, E.E. and Lambert, S.J. (eds.) *Work and Life Integration.* Mahwah, NJ: Lawrence Erlbaum, 349–374.

Lewis, S., Gambles, R. and Rapoport, R. 2007. The constraints of a 'work-life balance' approach: an international perspective. *International Journal of Human Resource Management* 18(3): 360–373.

Lewis, S., Rapoport, R. and Gambles, R. 2003. Reflections on the integration of paid work with the rest of life. *Journal of Managerial Psychology* 18(8): 824–841.

Lier, D.C. 2007. Places of work, scales of organising: a review of labour geography. *Geography Compass* 1(4): 814–833.

Lindberg, L.D. 1996. Women's decisions about breastfeeding and maternal employment. *Journal of Marriage and the Family* 58(1): 239–251.

Lundvall, B. 1992. *National Systems of Innovation.* London: Frances Pinter.

Lundvall, B.Ä. and Johnson, B. 1994. The learning economy. *Journal of Industry Studies* 1(2): 23–42.

MacDonald, M., Phipps, S. and Lethbridge, L. 2005. Taking its toll: the influence of paid and unpaid work on women's well-being. *Feminist Economics* 11(1): 63–94.

MacEwen, K.E. and Barling, J. 1994. Daily consequences of work interference with family and family interference with work. *Work and Stress* 8: 244–254.

MacKinnon, D., Cumbers, A. and Chapman, K. 2002. Learning, innovation and regional development: a critical appraisal of recent debates. *Progress in Human Geography* 26(3): 293–311.

Malecki, E.J. and Oinas, P. (eds.) 1999. *Making Connections.* Aldershot: Ashgate.

Malmberg, A. and Maskell, P. 2002. The elusive concept of localization economies: towards a knowledge-based theory of spatial clustering. *Environment and Planning A* 34: 429–449.

Markham, W.T., Macken, P.O., Bonjean, C.M. and Corder, J. 1983. A note on sex, geographic mobility and career advancement. *Social Forces* 1138–1146.

Markusen, A. 1996. Sticky places in slippery space: a typology of industrial districts. *Economic Geography* 72(3): 293–313.

Markusen, A. 1999. Fuzzy concepts, scanty evidence, policy distance: the case for rigour and policy relevance in critical regional studies. *Regional Studies* 33(9): 869–884.

Marmot, M., Allen, J., Goldblatt, P., Boyce, T., McNeish, D., Grady, M. and Geddes, I. 2010. *Fair Society, Healthy Lives: The Marmot Review.* London: University College London.

Marshall, A. [1890] 2013. *Principles of Economics*, 8th edition. London: Macmillan.

Marshall, B.L. 1994. *Engendering Modernity.* Cambridge: Polity Press.

Martin, H.B. and MacDonnell, R. 2012. Is telework effective for organizations? *Management Research Review* 35(7): 602–616.

Martin, R. 2013. Differentiated knowledge bases and the nature of innovation networks. *European Planning Studies* 21(9): 1418–1436.

Martin, R. and Moodysson, J. 2013. Comparing knowledge bases: on the geography and organization of knowledge sourcing in the regional innovation system of Scania, Sweden. *European Urban and Regional Studies* 20(2): 170–187.

Martin, R.L. 2000. Institutionalist approaches in economic geography. In Sheppard, E. and Barnes, T.J. (eds.) *A Companion to Economic Geography*. Oxford: Blackwell, 77–94.

Martin, R.L. 2006. *The Competitive Performance of English Cities: Competitiveness Case Study of Cambridge*. London: Office of the Deputy Prime Minister.

Martin, R.L. and Sunley, P. 2003. Deconstructing clusters: chaotic concept or policy panacea? *Journal of Economic Geography* 3(1): 5–35.

Martínez Sánchez, A., Pérez Pérez, M., de Luis Carnicer, P. and José Vela Jiménez, M. 2007. Teleworking and workplace flexibility: a study of impact on firm performance. *Personnel Review* 36(1): 42–64.

Maskell, P. 2001. Towards a knowledge-based theory of the geographical cluster. *Industrial and Corporate Change* 10(4): 921–943.

Maskell, P. and Malmberg, A. 1999. Localised learning and industrial competitiveness. *Cambridge Journal of Economics* 23(2): 167–185.

Mason, R. 2003. Listening to lone mothers: paid work, family life, and childcare in Canada. *Journal of Children and Poverty* 9(1): 41–54.

Massey, D. 1995. Masculinity, dualisms and high technology. *Transactions of the Institute of British Geographers* 20: 487–499.

Massey, D. 1997. Economic/non-economic. In Lee, R. and Wills, J. (eds.) *Geographies of Economies*. London: Arnold, 27–36.

Mathews, G. 2003. Can 'a real man' live for his family: Ikigai and masculinity in today's Japan. In Roberson, J.E. and Suzuki, N. (eds.) *Men and Masculinities in Contemporary Japan*. London: Routledge Curzon, 109–125.

Matthews, G. 2003. *Silicon Valley, Women, and the Californian Dream*. Stanford, CA: Stanford University Press.

Maule, A.J. and Edland, A.C. 1997. The effects of time pressure on human judgment and decision making. In Ranyard, R., Crozier, W.R. and Svenson, O. (eds.) *Decision Making: Cognitive Models and Explanations*. London: Routledge, 189–204.

Maxwell, G.A. and McDougall, M. 2004. Work-life balance: exploring the connections between levels of influence in the UK public sector. *Public Management Review* 6(3): 377–393.

Maxwell, G., Rankine, L., Bell, S. and MacVicar, A. 2007. The incidence and impact of flexible working arrangements in smaller businesses. *Employee Relations* 29(2): 138–161.

Mayer, H. 2013. Entrepreneurship in a hub-and-spoke industrial district: firm survey evidence from Seattle's technology industry. *Regional Studies* 47(10): 1715–1733.

McCampbell, A.S. 1996. Benefits achieved through alternative work schedules. *Human Resource Planning* 19(3): 133–146.

McCann, E.J. 2007. Inequality and politics in the creative-city region. *International Journal of Urban and Regional Research* 31(1): 188–196.

McCann, P. and Simonen, J. 2005. Innovation, knowledge spillovers and local labour markets. *Papers in Regional Science* 84(3): 465–485.

McDowell, L.M. 1991. Life without father and Ford: the new gender order of post-Fordism. *Transactions of the Institute of British Geographers* 16: 400–419.

McDowell, L.M. 1997. *Capital Culture: Gender at Work in the City.* Oxford: Blackwell.

McDowell, L.M. 2000. Feminists rethink the economic: the economics of gender, the gender of economics. In Clark, G.L., Feldman, M.P. and Gertler, M.S. (eds.) *The Oxford Handbook of Economic Geography.* Oxford: Oxford University Press, 497–517.

McDowell, L.M. 2004. Work, workfare, work/life balance and an ethic of care. *Progress in Human Geography* 28(2): 145–163.

McDowell, L.M. 2010. Capital culture revisited: sex, testosterone and the city. *International Journal of Urban and Regional Research* 34(3): 652–658.

McDowell, L.M. 2014a. The lives of others: body work, the production of difference, and labor geographies. *Economic Geography* 91(1): 1–23.

McDowell, L.M. 2014b. Gender, work, employment and society: feminist reflections on continuity and change. *Work, Employment and Society* 28(5): 825–837.

McDowell, L.M. and Dyson, J. 2011. The other side of the knowledge economy: 'reproductive' employment and affective labours in Oxford. *Environment and Planning A* 43(9): 2186–2201.

McDowell, L.M., Perrons, D., Fagan, C., Ray, K. and Ward, K. 2005. The contradictions and the intersections of class and gender in a global city: placing working women's lives on the research agenda. *Environment and Planning A* 37(3): 441–467.

McDowell, L., Ward, K., Fagan, C., Perrons, D. and Ray, K. 2006. Connecting time and space: the significance of transformations in women's work in the city. *International Journal of Urban and Regional Research* 30(1): 141–158.

McGinnity, F. and Russell, H. 2007. Work rich, time poor? Time-use of women and men in Ireland. *Economic and Social Review* 38(3): 323–354.

McGinnity, F., Russell, H. and Smyth, E. 2008. Gender, work-life balance and quality of life. In Fahey, T., Russell, H. and Whelan, C.T. (eds.) *Quality of Life in Ireland.* Milton Keynes: Springer, 199–215.

McGovern, P., Hill, S., Mills, C. and White, M. 2007. *Market, Class and Employment.* Oxford: Oxford University Press.

McKie, L., Gregory, S. and Bowlby, S. 2002. Shadow times: the temporal and spatial frameworks and experiences of caring and working. *Sociology* 36: 897–924.

McNall, L.A., Masuda, A.D. and Nicklin, J.M. 2009. Flexible work arrangements, job satisfaction, and turnover intentions. *Journal of Psychology* 144(1): 61–81.

Melymuka, K. 2000. IT women in Silicon Valley. *Computerworld Inc.* 13 March.

Miller, H.T. and Fox, C.J. 2001. The epistemic community. *Administration and Society* 32(6): 668–685.

Milliken, F.J. and Martins, L.L. 1996. Searching for common threads: understanding the multiple effects of diversity in organizational groups. *Academy of Management Review* 21(2): 402–433.

Minnotte, K.L. 2012. Family structure, gender, and the work–family interface: work-to-family conflict among single and partnered parents. *Journal of Family and Economic Issues* 33(1): 95–107.

Mitchell, K., Marston, S.A. and Katz, C. 2004. *Life's Work.* Oxford: Blackwell.

Mitchell, T.R., Holtom, B.C. and Lee, T.W. 2001. How to keep your best employees. *Academy of Management Executive* 15(4): 96–108.

Moen, P. (ed.) 2003. *It's About Time: Couples and Careers*. Ithaca, NY: Cornell University Press.

Moen, P. and Sweet, S. 2004. From 'work–family' to 'flexible careers': a life course reframing. *Community, Work and Family* 7(2): 209–226.

Moody, F. 1995. *I Sing the Body Electronic*. London: Hodder and Stoughton.

Moore, J.E. 2000. One road to turnover: an examination of work exhaustion in technology professionals. *MIS Quarterly* 24(1): 141–168.

Morgan, K. 2004. Sustainable regions: governance, innovation and scale. *European Planning Studies* 12(6): 871–889.

Mumford, M.D. 2000. Managing creative people: strategies and tactics for innovation. *Human Resource Management Review* 10(3): 313–351.

NCWIT 2014. *National Centre for Women in Information Technology Scorecard*. Boulder, CO: University of Colorado.

Negrey, C. 2012. *Work Time: Conflict, Control and Change*. Cambridge: Polity Press.

Nelson, J.A. 1992. Gender, metaphor, and the definition of economics. *Economics and Philosophy* 8(1): 103–125.

Nelson, J.A. 1995. Feminism and economics. *Journal of Economic Perspectives* 9(2): 131–148.

Nelson, R. and Winter, S.G. 1982. *Evolutionary Theory of Economic Change*. Cambridge, MA: Harvard University Press.

Nemeth, C. 1986. Differential contributions of majority and minority influence. *Psychological Review* 93: 23–32.

Newell, S., Robertson, M., Scarbrough, H. and Swan, J. 2002. *Managing Knowledge Work*. Basingstoke: Palgrave.

New York Times, The 2013. Bank of America intern in London dies. 20 August.

New York Times, The 2015. Reflections on stress and long hours on Wall Street. 1 June.

Nohria, N. and Eccles, R. 1992. *Networks and Organizations*. Boston: Harvard Business School Press.

Nonaka, I. and Takeuchi, H. 1995. *The Knowledge Creating Company*. New York: Oxford University Press.

Noor, S. and Maad, N. 2009. Examining the relationship between work life conflict, stress and turnover intentions among marketing executives in Pakistan. *International Journal of Business and Management* 3(11): 93–102.

O'Boyle, C.A. 2006. Getting work-life balance right. Invited key-note address. 3rd Occupational Health and Safety Summit, Dublin, 2 October.

O'Brien, M. 2005. *Shared Caring: Bringing Fathers into the Frame*. Manchester: Equal Opportunities Commission (working paper series).

O'Callaghan, C., Kelly, S., Boyle, M. and Kitchin, R. 2015. Topologies and topographies of Ireland's neoliberal crisis. *Space and Polity* 19(1): 31–46.

O'Carroll, A. 2005. The long and the short of it: working time in the Irish IT sector. In Boucher, G. and Collins, G. (eds.) *The New World of Work*. Dublin: The Liffey Press, 163–180.

O'Connell, P.J. 1999. Sick man or tigress? The labour market in the Republic of Ireland. In Heath, A.F., Breen, R. and Whelan, C.T. (eds.) *Ireland North and South*. Oxford: Oxford University Press, 215–249.

O'Connell, P.J. and Russell, H. 2005. *Equality at Work?* Dublin: The Equality Authority.

O'Connell, P.J., Russell, H., Williams, J. and Blackwell, S. 2003. *The Changing Workplace*. Dublin: NCPP.

O'Connell, P.J., Russell, H., Watson, D. and Byrne, D. 2010. *The Changing Workplace*. Dublin: NCPP.

OECD 1996. *Technology, Productivity, and Job Creation: Best Policy Practices*. Paris: OECD.

OECD 2003. *Babies and Bosses: Reconciling Work and Family Life – Austria, Ireland and Japan*. Paris: OECD.

OECD 2005. *Babies and Bosses: Reconciling Work and Family Life – Canada, Finland, Sweden and the United Kingdom*. Paris: OECD.

OECD 2007a. *Babies and Bosses: Reconciling Work and Family Life – A Synthesis of Findings for OECD Countries*. Paris: OECD.

OECD 2007b. *Innovation and Growth: Rationale for an Innovation Strategy*. Paris: OECD.

OECD 2012. *Closing the Gender Gap – Act Now*. Paris: OECD.

OECD 2014. *All On Board: Making Inclusive Growth Happen*. Paris: OECD.

Oinas, P. and Malecki, E.J. 2002. The evolution of technologies in time and space: from national and regional to spatial innovation systems. *International Regional Science Review* 25(1): 102–131.

Oldham, G.R. and Cummings, A. 1996. Employee creativity: personal and contextual factors at work. *Academy of Management Journal* 39(3): 607–634.

Ong, H.L.C. and Jeyaraj, S. 2014. Work-life interventions: differences between work-life balance and work-life harmony and its impact on creativity at work. *Sage Open* 2014(4): 1–11.

ONS 2013. *Women in the Labour Market*. London: UK Office for National Statistics.

O'Reilly, C.A. 1989. Corporations, culture and commitment: motivation and social control in organizations. *California Management Review*, Summer, 9–25.

Ó Riain, S. 1997. An offshore Silicon Valley? The emerging Irish software industry. *Competition and Change* 2: 175–212.

Ó Riain, S. 2000. Net-working for a living: Irish software developers in the global workplace. In Burawoy, M., Blum, J.A., George, S., Gille, Z. and Thayer, M. (eds.) *Global Ethnography*. Berkeley: University of California Press, 175–202.

Ó Riain, S. 2004. The politics of mobility in technology-driven commodity chains: developmental coalitions in the Irish software industry. *International Journal of Urban and Regional Research* 28(3): 642–663.

Ó Riain, S. 2014. *The Rise and Fall of Ireland's Celtic Tiger*. Cambridge: Cambridge University Press.

Östbring, L. and Lindgren, U. 2013. Labour mobility and plant performance. *Geografiska Annaler B* 95(4): 287–305.

Osterlow, M. and Frey, B.S. 2000. Motivation, knowledge transfer and organizational forms. *Organization Science* 11(5): 538–550.

O'Toole, J. 1974. *Work and the Quality of Life*. Cambridge, MA: MIT Press.

Panteli, A., Stack, J., Atkinson, M. and Ramsay, H. 1999. The status of women in the UK IT industry. *European Journal of Information Systems* 8(3): 170–182.

Panteli, N. and Pen, S. 2007. Empowering women returners in the UK high-tech industry. *Personnel Review* 39(1): 44–61.

Panteli, N., Stack, J. and Ramsey, H. 2001. Gendered patterns in computing work in the late 1990s. *New Technology, Work and Employment* 16(1): 3–17.

Payne, G. and Abbott, P. (eds.) 1990. *The Social Mobility of Women: Beyond Male Mobility Models*. Basingstoke: Falmer Press.

Peck, J. 1996. *Work-Place*. London: Guilford Press.

Pelled, L.H., Eisenhardt, K.M. and Xin, K.R. 1999. Exploring the black box: an analysis of work group diversity, conflict and productivity. *Administrative Science Quarterly* 44: 1–28.

Perlow, L.A. 1997. *Finding Time: How Corporations, Individuals, and Families Can Benefit from New Work Practices*. Ithaca, NY: Cornell University Press.

Perlow, L.A. 1999. The time famine: toward a sociology of work time. *Administrative Science Quarterly* 44(1): 57–81.

Perrons, D. 2001. Towards a more holistic framework for economic geography. *Antipode* 33(3): 208–215.

Perrons, D. 2003. The new economy and the work-life balance: conceptual explorations and a case study of new media. *Gender, Work and Organization* 10(1): 65–93.

Perrons, D. 2004. Equity and representation in the new economy. In Kelly, J. and Willman, P. (eds.) *Union Organization and Activity*. London: Routledge.

Perrons, D. 2010. Gender, work and 'market' values. *Renewal: A Journal of Social Democracy* 18: 34–42.

Perrons, D. 2012. Regional performance and inequality: linking economic and social development through a capabilities approach. *Cambridge Journal of Regions, Economy and Society* 5(1): 15–29.

Perrons, D. and Dunford, M. 2013. Regional development, equality and gender: moving towards more inclusive and socially sustainable measures. *Economic and Industrial Democracy* 34(3): 483–499.

Perrons, D., Fagan, C., McDowell, L. and Ward, K. (eds.) 2006. *Gender Divisions and Working Time in the New Economy*. Cheltenham: Edward Elgar.

Perrons, D., Plomien, A. and Kilkey, M. 2010. Migration and uneven development within an enlarged European Union: fathering, gender divisions and male migrant domestic services. *European Urban and Regional Studies* 17(2).

Perry-Smith, J. and Blum, T. 2000. Work-family human resource bundles and perceived organizational performance. *Academy of Management Journal* 43(6): 1107–1117.

Peterson, H. 2007. Gendered work ideals in Swedish IT firms. *Gender, Work and Organization* 14(4): 333–348.

Pfeffer, J. 1994. *Competitive Advantage Through People*. Boston: Harvard Business School Press.

Phillips, J., Bernard, M. and Chittenden, M. 2002. *Juggling Work and Care: The Experiences of Working Carers of Older Adults*. Oxford: Policy Press.

Pike, A., Rodríguez-Pose, A. and Tomaney, J. 2006. *Local and Regional Development*. Abingdon: Routledge.

Pinch, S., Henry, N., Jenkins, M. and Tallman, S. 2003. From 'industrial districts' to 'knowledge clusters': a model of knowledge dissemination and competitive advantage in industrial agglomerations. *Journal of Economic Geography* 3(4): 373–388.

Pittman, J.F., Solheim, C.A. and Blanchard, D. 1996. Stress as a driver of the allocation of housework. *Journal of Marriage and the Family* 58(2): 456–468.

Pixley, J.E. 2008. Life course patterns of career-prioritizing decisions and occupational attainment in dual-earner couples. *Work and Occupations* 35(2): 127–163.

Pocock, B., Skinner, N. and Williams, P. 2008. Work-life outcomes in Australia: concepts, outcomes and policy. In Warhurst, C., Eikhof, D.R. and Haunschild, A. (eds.) *Work Less, Live More?* Basingstoke: Palgrave Macmillan, 22–43.

Polanyi, M. 1967. *The Tacit Dimension.* London: Routledge and Kegan Paul.

Pollard, J. 2012. Gendering capital: financial crisis, financialization and (an agenda for) economic geography. *Progress in Human Geography* 37(3): 403–423.

Porter, M.E. 1985. *Competitive Advantage.* New York: Free Press.

Porter, M.E. 1998. *On Competition.* Cambridge, MA: Harvard University Press.

Porter, S. and Ayman, R. 2010. Work flexibility as a mediator of the relationship between work–family conflict and intention to quit. *Journal of Management and Organization* 16(3): 411–424.

Power, D. and Lundmark, M. 2004. Working through knowledge pools: labour market dynamics, the transference of knowledge and ideas, and industrial clusters. *Urban Studies* 41(5/6): 1025–1044.

Presser, H. 2006. Employment in a 24/7 economy: challenges for the economy. In Perrons, D., Fagan, C., McDowell, L. and Ward, K. (eds.) *Gender Divisions and Working Time in the New Economy.* Cheltenham: Edward Elgar, 35–57.

PricewaterhouseCoopers (PWC) and Demos 2013. *Good Growth for Cities.* London: PWC and Demos.

Ransom, C. and Burud, S. 1989. *Productivity Impact Study of an On-Site Child Care Center.* Burud and Associates, Pasadena, CA.

Ransome, P. 2007. Conceptualizing boundaries between 'life' and 'work'. *International Journal of Human Resource Management* 18(3): 374–386.

Ranson, G. 2010. *Against the Grain: Couples, Gender and the Reframing of Parenting.* Toronto: University of Toronto Press.

Rapoport, R. and Rapoport, R. 1965. Work and family in contemporary society. *American Sociological Review* 30(3): 381–394.

Rapoport, R., Bailyn, L., Fletcher, J. and Pruitt, B. 2002. *Beyond Work-Family Balance: Advancing Gender Equity and Workplace Performance.* London: Jossey-Bass.

Reagans, R. and Zuckerman, E.W. 2001. Networks, diversity, and productivity: the social capital of corporate R&D teams. *Organization Science* 12(4): 502–517.

Redmond, J., Valiulis, M. and Drew, E. 2006. *Literature Review of Issues Related to Work-Life Balance, Workplace Culture and Maternity/Childcare Issues.* Dublin: Trinity College Dublin.

Redmond, M.R., Mumford, M.D. and Tech, R.J. 1993. Putting creativity to work: leader influences on subordinate creativity. *Organizational Behavior and Human Decision Processes* 55: 120–151.

Rees, T. 2000. The learning region? Integrating gender equality into regional economic development. *Policy and Politics* 28(2): 179–191.

Rifkin, J. 2014. *The Zero Marginal Cost Society.* Basingstoke: Palgrave Macmillan.

Roberts, K. 2007. Work-life balance – the sources of the contemporary problem and probable outcomes: a review and interpretation of the evidence. *Employee Relations* 29(4): 334–351.

Robinson, G. and Dechant, K. 1997. Building a business case for diversity. *Academy of Management Executive* 11(3): 21–31.

Roche, F., O'Shea, R., Allen, T.J. and Breznitz, D. 2008. The dynamics of an emerging entrepreneurial region in Ireland. In Phan, P.H.C., Venkataraman, S. and Velamuri, S.R. (eds.) *Entrepreneurship in Emerging Regions Around the World.* Cheltenham: Edward Elgar, 9–46.

Rochlen, A., Suizzo, M., McKelley, R. and Scaringi, V. 2008. 'I'm just providing for my family'. *Psychology of Men and Masculinity* 9(4): 193–206.

Rodgers, C. 1992. The flexible workplace: what have we learned? *Human Resource Management* 31(3): 183–199.

Rodríguez-Pose, A. and Wilkie, C. 2015. *Conceptualizing Equitable Economic Growth in Urban Environments*. London: LSE Cities Alliance.

Roper, S. and Grimes, S. 2005. Wireless Valley, Silicon Wadi and Digital Island – Helsinki, Tel Aviv and Dublin in the ICT global production network. *Geoforum* 36: 297–313.

Rubery, J. 1997. What do women want from full employment? In Philpott, J. (ed.) *Working for Full Employment*. London: Routledge, 63–80.

Rubery, J. and Rafferty, A. 2013. Women and recession revisited. *Work, Employment and Society* 27(3): 414–432.

Rubery, J., Earnshaw, J., Marchington, M., Cooke, F.L. and Vincent, S. 2002. Changing organizational forms and the employment relationship. *Journal of Management Studies* 39(5): 645–672.

Russell, H. and McGinnity, F. 2014. Under pressure: the impact of recession on employees in Ireland. *British Journal of Industrial Relations* 52(2): 286–307.

Russell, H., Layte, R., Maitre, B., O'Connell, P. and Whelan, C. 2004. *Work-Poor Households*. ESRI Policy Research Series No. 52. Dublin, Economic and Social Research Institute.

Russell, H., O'Connell, P.J. and McGinnity, F. 2009. The impact of flexible working arrangements on work-life conflict and work pressure in Ireland. *Gender, Work and Organization* 16(1): 73–97.

Rutherford, S. 2001. 'Are you going home already?' The long hours culture, women managers and patriarchal closure. *Time and Society* 10(2/3): 259–276.

Rutherford, T. 2010. De/re-centring work and class? A review and critique of labour geography. *Geography Compass* 4(7): 768–777.

Saxenian, A. 1994. *Regional Advantage: Culture and Competition in Silicon Valley and Route 128*. Cambridge, MA: Harvard University Press.

Sayah, S. 2013. Managing work–life boundaries with information and communication technologies: the case of independent contractors. *New Technology, Work and Employment* 28(3): 179–196.

Scandura, T.A. and Lankau, M.J. 1997. Relationships of gender, family responsibility and flexible work hours to organizational commitment and job satisfaction. *Journal of Organizational Behavior* 18: 377–391.

Scase, R. and Scales, J. 1998. *Work Now – Pay Later?* Swindon: ESRC.

Schoenberger, E. 1997. *The Cultural Crisis of the Firm*. Oxford: Blackwell.

Schoenberger, E. 2001. Interdisciplinarity and social power. *Progress in Human Geography* 25(3): 365–382.

Scholarios, D. and Marks, A. 2004. Work-life balance and the software worker. *Human Resource Management Journal* 14(2): 54–74.

Schor, J.B. 1991. *The Overworked American*. New York: Basic Books.

Scott, A.J. 1988. *New Industrial Spaces*. London: Pion.

Scott, A.J. 2000. Economic geography: the great half century. In Clark, R.L., Feldman, M. and Gertler, M. (eds.) *Handbook of Economic Geography*. Oxford: Oxford University Press, 18–44.

Scott, A.J. and Storper, M. 1987. High technology industry and regional development: a theoretical critique and reconstruction. *International Social Science Journal* 112: 215–232.

Scott, J.L., Crompton, R. and Lyonette, C. (eds.) 2010. *Gender Inequalities in the 21st Century*. Cheltenham: Edward Elgar.

Segal Quince Wicksteed. 1985. *The Cambridge Phenomenon: The Growth of High Technology Industry in a University Town*. Cambridge: SQW.

Sen, A. 1999. *Development as Freedom*. Oxford: Oxford University Press.

Sennett, R. 1998. *The Corrosion of Character*. London: Norton.

Sennett, R. 2006. *The Culture of the New Capitalism*. New Haven, CT and London: Yale University Press.

Sexton, J.J., Hughes, G. and Finn, C. 2002. *Occupational Employment Forecasts 2015*. Dublin: Economic and Social Research Institute.

Seylor, D., Monroe, P. and Garand, J. 1993. Executives' assessment of company sponsored family benefits. *Family Perspectives* 27(2): 147–162.

Sharone, O. 2002. *Engineering Consent: Overwork and Anxiety at a High-Tech Firm*. Berkeley, CA: Center for Working Families.

Shepard, E.M., Clifton, T.J. and Kruse, D. 1996. Flexible work hours and productivity: some evidence from the pharmaceutical industry. *Industrial Relations* 35(1): 123–139.

Shih, J. 2004. Project time in Silicon Valley. *Qualitative Sociology* 27(2): 223–245.

Shih, J. 2006. Circumventing discrimination: gender and ethnic strategies in Silicon Valley. *Gender and Society* 20(2): 177–206.

Silicon Valley Community Foundation 2015. *Silicon Valley Competitiveness and Innovation Project 2015*. Palo Alto, CA: SVCF with the Silicon Valley Leadership Group.

SIPTU/Callender, R. 2003. Balancing work and life on the back of a limping Celtic Tiger. Paper presented to the 'Improving Life At Work' conference, London, January.

Sirianni, C. and Negrey, C. 2000. Working time as gendered time. *Feminist Economics* 6(1): 59–76.

Smithson, J. and Stokoe, E.H. 2005. Discourses of work-life balance: negotiating 'gender-blind' terms in organizations. *Gender, Work and Organization* 12(2): 147–168.

Snyder, K. 2014. Why women leave tech – it's the culture, not because 'math is hard'. *Fortune Magazine*, 2 October.

Song, J., Almeida, P. and Wu, G. 2003. Learning-by-hiring. *Management Science* 49(4): 351–365.

Song, Y. and Knaap, G. J. 2004. Measuring urban form: is Portland winning the war on sprawl? *Journal of the American Planning Association* 70(2): 210–225.

Spender, D. 1980. *Man Made Language*. London: Routledge and Kegan Paul.

Spender, J.C. 1996. Making knowledge the basis for a dynamic theory of the firm. *Strategic Management Journal* 17: 45–62.

Staeheli, L. 2013. Whose responsibility is it? Obligation, citizenship and social welfare. *Antipode* 45(3): 521–540.

Stalk, G. and Haut, T.M. 1990. *Competing Against Time*. New York: Free Press.

Stangel, L. 2013. Facebook's 12 most fantastic employee perks. *Silicon Valley Business Journal*, April 8.

Stanworth, C. 2000. Women and work in the information age. *Gender, Work and Organization* 7(1): 20–32.

Stedman Jones, G. 2014. *Outcast London*. London: Verso.

Sterne, J. 2004. *Adventures in Code: The Story of the Irish Software Industry*. Dublin: The Liffey Press.

St John's Innovation Centre 2013. *Cambridge Technopole Report 2013*. Cambridge: St John's Innovation Centre.

Storper, M.J. 1995. The resurgence of regional economies, ten years later. *European Urban and Regional Studies* 2: 191–222.

Storper, M.J. 1997. *The Regional World*. New York: Guilford Press.

Storper, M.J. and Venables, A.J. 2002. Buzz: face-to-face contact and the urban economy. *Journal of Economic Geography* 4(4): 351–370.

Strazdins, L. and Broom, D.H. 2004. Acts of love (and work): gender imbalance in emotional work and women's psychological distress. *Journal of Family Issues* 25(3): 356–378.

Sullivan, C. and Lewis, S. 2001. Home-based telework, gender, and the synchronization of work and family. *Gender, Work and Organization* 8(2): 123–145.

Swan, J., Morris, L., Werner, S. and Jackson, S. 2011. *Top Employers For Working Families*. London: Working Families.

Tang, N. and Cousins, C. 2005. Working time, gender and family: an east–west European comparison. *Gender, Work and Organization* 12(6): 527–550.

Tausig, M. and Fenwick, R. 2001. Unbinding time: alternate work schedules and work-life balance. *Journal of Family and Economic Issues* 22(2): 101–119.

Taylor, P. 2013. *Performance Management and the New Workplace Tyranny*. Glasgow: University of Strathclyde.

Taylor, R. 2001. *The Future of Work-Life Balance*. Swindon: ESRC.

Tesluk, P.E., Farr, J.L. and Klein, S.R. 1997. Influences of organizational culture and climate on individual creativity. *Journal of Creative Behavior* 31(1): 27–41.

Thompson, L.F. and Aspinwall, K.R. 2009. The recruitment value of work/life benefits. *Personnel Review* 38(2): 195–210.

Thrift, N.J. and Olds, K. 1996. Refiguring the economic in economic geography. *Progress in Human Geography* 20(3): 311–337.

Thulin, P. 2009. *Labor mobility, knowledge diffusion and regional growth*. Stockholm, Sweden: Centre of Excellence for Science and Innovation Studies (Working Paper 209).

Tipping, S., Chanfreau, J., Perry, J. and Tait, C. 2012. *The Fourth Work-Life Balance Employee Survey*. London: Department for Business Innovation and Skills.

Titze, M., Brachert, M. and Kubis, A. 2014. Actors and interactions – identifying the role of industrial clusters for regional production and knowledge generation activities. *Growth and Change* 45(20): 163–190.

Tomaney, J., Pike, A. and Rodriguez-Pose, A. 2010. Local and regional development in times of crisis. *Environment and Planning A* 42: 771–779.

Tran, M. 2014. Apple and Facebook offer to freeze eggs for female employees. *The Guardian*, 15 October.

Trippl, M. 2013. Scientific mobility and knowledge transfer at the interregional and intraregional level. *Regional Studies* 47(10): 1653–1667.

TUC 2005. *Work-Life Balance*. London: Trades Union Congress.

TUC 2009. *Women and Recession*. London: Trades Union Congress.

TUC 2012. *TUC Equality Audit 2012*. London: Trades Union Congress.

TUC 2015a. *The Impact on Women of Recession and Austerity*. London: Trades Union Congress.

TUC 2015b. *Workers Contribute £32bn to UK Economy from Unpaid Overtime*. London: Trades Union Congress.

Tyree, A. and Treas, J. 1974. The occupational and marital mobility of women. *American Sociological Review* 39(3): 293–302.

Upadhya, C. and Vasavi, A.R. 2006. *Work, Culture and Sociality in the Indian Information Technology (IT) Industry: A Sociological Study*. Bangalore: National Institute of Advanced Studies.

US Bureau of Labor Statistics 2005. *Women in the Labor Force: A Databook*. Washington, DC: US Department of Labor.

US Census Bureau 2006. *2002 Economic Census: Survey of Business Owners Company Statistics Series*. Washington, DC: US Department of Commerce.

US Council of Economic Advisers 2010. *Work-Life Balance and the Economics of Workplace Flexibility*. Washington, DC: Executive Office of the President, Council of Economic Advisers. Available at www.worldatwork.org/adimLink?id=50354 (accessed 15 March 2017).

Valcour, P.M. and Tolbert, P. 2003. Gender, family and career in the era of boundarylessness. *International Journal of Human Resource Management* 14(5): 768–787.

Van Staveren, I., Elson, D., Grown, C. and Cagatay, N. (eds.) 2012. *Feminist Economics of Trade*. Abingdon: Routledge.

Van Wanrooy, B., Bewley, H., Bryson, A., Forth, J., Freeth, S., Stokes, L. and Wood, S. 2013. *Employment Relations in the Shadow of Recession*. Basingstoke: Palgrave Macmillan.

Vosberg, S.K. 1998. The effects of positive and negative affect on divergent thinking performance. *Creativity Research Journal* 11: 165–172.

Voydanoff, P. 1988. Work role characteristics, family structure demands, and work/family conflict. *Journal of Marriage and the Family* 50: 749–761.

Wajcman, J. 1991. *Feminism Confronts Technology*. Cambridge: Polity Press.

Wajcman, J. 1999. *Women and Men in Corporate Management*. Cambridge: Allen & Unwin.

Walby, S. 1986. *Gender at Work*. Cambridge: Polity Press.

Walby, S. 1997. *Gender Transformations*. London: Routledge.

Waldfogel, J., Higuchi, Y. and Abe, M. 1999. Family leave policies and women's retention after childbirth. *Journal of Population Economics* 12(4): 523–545.

Walkowitz, D.J. 1978. *Worker City, Company Town*. Chicago: University of Illinois Press.

Wayne, J. H. and Cordeiro, B.L. 2003. Who is a good organizational citizen? Social perception of male and female employees who use family leave. *Sex Roles* 49: 233–246.

Webgrrls 2007. *Webgrrls User Survey 2007*. Retrieved 16 October 2007 from www.webgrrls.com/about/MediaKit.pdf

Weeks, K. 2011. *The Problem with Work*. London: Duke University Press.

West, M.A. and Sacramento, C.A. 2007. Flourishing in teams: developing creativity and innovation. In Henry, J. (ed.) *Creative Management and Development*, 3rd edition. London: Sage, 25–44.

White, M., Hill, S., McGovern, P., Mills, C. and Smeaton, D. 2003. 'High-performance' management practices, working hours and work-life balance. *British Journal of Industrial Relations* 41(2): 175–195.

Whitehouse, G., Haynes, M., McDonald, F. and Arts, D. 2007. *Reassessing the Family-Friendly Workplace*. London: Department for Business, Enterprise and Regulatory Reform.

Williams, J. 2000. *Unbending Gender*. Oxford: Oxford University Press.

Williams, P., Pocock, B. and Skinner, N. 2008. 'Clawing back time': expansive working time and implications for work–life outcomes in Australian workers. *Work, Employment and Society* 22(4): 737–748.

Williams, Z. 2014. Freezing women's eggs? The tech industry isn't modern, it's Neanderthal. *The Guardian*, 15 October.

Wills, J., Datta, K., Evans, Y., Herbert, J., May, J. and McIlwaine, C. 2010. *Global Cities at Work*. London: Pluto.

Wilson, F. 2003. Can compute, won't compute: women's participation in the culture of computing. *New Technology, Work and Employment* 18(2): 127–142.

Wilson, L. and Spoehr, J. 2010. Labour relations and the transfer of knowledge in industrial clusters. *Geographical Research* 48: 42–51.

Wise, S. 2003. *Work Life Balance: Literature and Research Review*. London: DTI and Fair Play.

Wise, S. and Bond, S. 2003. Work-life balance: does it do exactly what it says on the tin? *Women in Management Review* 18(1/2): 20–31.

Wolfe, D.A. and Gertler, M.S. 2004. Clusters from the inside and out: local dynamics and global linkages. *Urban Studies* 41: 1071–1093.

Wong, S.C.K. and Ko, A. 2009. Exploratory study of understanding hotel employees' perception on work–life balance issues. *International Journal of Hospitality Management* 28(2): 195–203.

Wood, S.J. and de Menezes, L.M. 2010. Family-friendly management, organizational performance and social legitimacy. *International Journal of Human Resource Management* 21(10): 1575–1597.

Working Families 2008a. *Flexible Working in a Challenging Economic Climate*. London: Working Families.

Working Families 2008b. *Flexible Working and Performance: Summary of Research*. London: Working Families.

World Economic Forum 2005. *Women's Empowerment: Measuring the Global Gender Gap*. Switzerland: World Economic Forum.

Wright, R. 1996. The occupational masculinity of computing. In Cheng, C. (ed.) *Masculinities in Organizations*. London: Sage, 77–96.

Yasbek, P. 2004. *The Business Case for Firm-Level Work-Life Balance Policies*. Wellington, New Zealand: Labour Market Policy Group, Department of Labour.

Yeung, H.W. 2005. Rethinking relational economic geography. *Transactions of the Institute of British Geographers* 30: 37–51.

Zachary, G.P. 1994. *Showstopper!* New York: Free Press.

Zucker, L.G. and Darby, M.R. 1996. Star scientists and institutional transformation: patterns of invention and innovation in the formation of the biotechnology industry. *Proceedings of the National Academy of Sciences* 93: 12709–12716.

Index

Work-Life Advantage: Sustaining Regional Learning and Innovation, First Edition. Al James.
© 2018 John Wiley & Sons Ltd. Published 2018 by John Wiley & Sons Ltd.